The Chemistry of Art

A guide to the pack

How to use this pack

This pack has been produced with secondary school science, in particular chemistry, in mind. The experiments in this pack could be used with students up to the age of 16, at the teacher's discretion. We have included a post-16 supplement, *The colour supplement,* of more advanced material for those who wish to take things further.

The 10 prints are all of paintings in the National Gallery, London, and have been selected because they illustrate a variety of different issues which are of interest and concern to both the scientists and the art historians at the Gallery. One print, *The Incredulity of Saint Thomas* by Cima, is also in the Salters' A-level Chemistry course. Another, *An Experiment on a Bird in the Air Pump* by Joseph Wright, has been chosen because it illustrates an early experiment, rather than for technical matters relating to the condition of the painting itself.

We suggest that, if you have room, you display the prints on laboratory walls, so that when work relating to them is done, the images are familiar. Your students may not like some of the images, and it is unlikely that they will see all of them as 'beautiful'; but we hope that they will come to find them interesting at some level.

Throughout the text of this pack you will find that words in **bold** are explained in the Glossary.

The Paintings

1) *Saint John the Baptist with Saint John the Evangelist and Saint James* by Nardo di Cione

2) *The Virgin and Child before a Firescreen* by a follower of Robert Campin

3) *Portrait of Alexander Mornauer* by the Master of the Mornauer Portrait.

4) *An Allegorical Figure* by Cosimo Tura

5) *The Incredulity of Saint Thomas* by Cima da Conegliano

6) *Bacchus and Ariadne* by Titian

7) *'The Stonemason's Yard'* by Canaletto

8) *An Experiment on a Bird in the Air Pump* by Joseph Wright

9) *Winter Landscape* by Caspar David Friedrich

10) *Boating on the Seine* by Pierre-Auguste Renoir

The experimental section

In this section there are suggestions for experiments and activities related to the work discussed in the paintings section. The experiments and activities are suitable for school and college laboratories and are designed to show how science, and in particular chemistry, is relevant to art. The material could also be used in science, chemistry or art clubs as extra-curricular activities, and as project ideas for the CREST awards scheme.

The colour supplement

This section is primarily for post-16 students and for teachers. However, interested pre-16 students will also benefit from reading the section.

The *Colour supplement* includes sections on the biology and chemistry of vision, the visible spectrum, the mixing of colours, and the causes of colour in objects. It ranges through aspects of biology and physics as well as chemistry and art. This may help to demonstrate – as does the rest of the pack – that the boundaries between the conventional 'subjects' are not facts of nature but are man-made.

About the National Gallery

The National Gallery building and its contents belong to the British nation. The Collection consists of about 2300 western European paintings dating from about 1250–1900 AD. It is possibly the finest and most comprehensive collection of its kind anywhere in the world, and contains numerous masterpieces by famous artists – *eg* Rembrandt, van Gogh and others – as well as first-rate works by less well known, or anonymous, painters. Major art galleries, such as the National Gallery, London, are important repositories of paintings which, through accidents of history, have survived to the present day. Each painting is unique and irreplaceable. Its survival is partly due to the skill of its maker, who would have been familiar with techniques of proven quality, and partly due to the care taken of it by a succession of previous owners. It is the responsibility of the present generation to ensure that these objects continue to survive. None of the paintings in the National Gallery was painted to hang there, and many have been damaged either through accident or by design – *eg* large paintings are sometimes reduced in size, and all are threatened by atmospheric pollution and ultraviolet radiation. The Gallery's prime purpose is to ensure that the paintings' condition remains stable and that no deterioration takes place. This is achieved by employing a small team of scientists and restorers, and it is due to their efforts that this pack exists.

Science at the National Gallery

The National Galley has a scientific department which employs six chemists. Their main concerns are:

- the analysis of pigments – these are largely inorganic chemicals;
- the analysis of media – organic chemicals; and
- preventative conservation – *eg* looking at issues of climate control and the effects of pollution.

Each of the scientists employed by the Gallery is a specialist in a particular area, such as organic, inorganic or analytical chemistry. Their work provides vital support for the work of the conservation department, which deals with the care of the paintings (both the side the public sees, and their backs). For example, when cleaning a painting the conservators need to know what materials they are dealing with, so that they can use appropriate chemicals. It is also important to know what pigments an artist used, for two reasons:

- to be able to distinguish between the artist's original work and later retouchings; and
- to ensure that any retouchings can be chemically different, so that future generations can distinguish between original paint and a restoration.

It is perhaps worth mentioning here that all retouchings are done using stable materials which can easily be removed again without affecting the original paint, and which do not themselves discolour.

Paintings (particularly those painted on wood panels) are very susceptible to changes of humidity. Modern air conditioning can be governed automatically by sensors that control the input of damp or dry air according to the prevailing conditions in a particular area.

Light damages all organic components of a painting, including the varnish and medium. It can also cause fading of certain pigments (particularly organic ones, such as lakes), and ultraviolet (UV) light is particularly harmful. The National Gallery filters out UV light.

Pollutants are monitored, and nitrogen oxides (such as NO_2) and sulfur dioxide (SO_2) are regarded as potentially damaging – not just to certain colours (ultramarine is the prime example, it discolours in acid conditions), but also to canvas supports. Air conditioning can filter out many of these pollutants.

Changes of temperature are not usually a problem to paintings, unless of course such changes are very extreme.

The National Gallery can sometimes offer free science tours to groups of school students. These take place in the Gallery in front of the paintings, and should be booked by telephone one term in advance. If you have timetabling problems why not book a joint visit with your art department?

For further details and bookings contact the education department by telephone on: 0207 747 2424; or by e-mail at: education@ng-london.org.uk

Information on all Gallery activities can be obtained from its website: www.nationalgallery.org.uk

The National Gallery
Trafalgar Square
London
WC2N 5DN

The Gallery regrets that visits to the Scientific or Conservation Departments are NEVER possible.

About the Royal Society of Chemistry

The Royal Society of Chemistry (RSC) is the professional body for chemists in the UK, and its Royal Charter objectives include the ensuring of an adequate supply of high quality chemists at all levels and a more chemically literate population. To these ends the Society produces a variety of materials to support and enhance the curriculum at all levels and for all age groups.

The RSC also produces careers materials for both students and teachers in a variety of formats. A fundamental message behind this material is that *Not all chemists wear white coats*, and that chemistry has relevance and use in many areas outside the chemical and pharmaceutical industries. The Society is also the largest non-governmental provider of in-service education and training for teachers of chemistry, running summer schools, industry study tours, hands-on symposia and other activities. It seeks to bring the relevance of chemistry to the widest possible audience by supporting the dissemination of the use of chemistry in novel contexts.

This publication, which has been written jointly by the RSC and the National Gallery, shows some of the links between chemistry and other subjects, and the relevance of chemistry to these subjects. It has direct relevance to both the art and the science curricula, and also provides additional material for teachers to use in the support of extra-curricular activities in both areas. It provides opportunities for both joint and independent project work within and between the art and science departments.

Further information about RSC educational activities can be obtained by writing to the

Education Department
Royal Society of Chemistry
Burlington House
Piccadilly
London W1V 0BN

(Tel 0207 437 8656)
(Fax 0207 287 9825)

or by e-mail: education@rsc.org

Full details of the activities of the Society can be found on the Society's website at www.rsc.org, and the Society is also the host to the website for chemistry societies and chemists worldwide at www.chemsoc.org. The latter site has links to other chemistry-related websites around the world.

Cloth of honour: a length of a fabric (usually cloth of gold or brocade) placed behind an important person to single them out.

Cobalt blue: a mixed cobalt-aluminium oxide, $CoO.Al_2O_3$. The French government sponsored research on the compounds of cobalt, with the intention of producing a better range of blues. The work of L.J.Thénard on the ores of cobalt led directly to cobalt blue.

Cobalt violet: a pure mauve pigment consisting of cobalt phosphate or cobalt arsenate, sometimes a mixture of both. Invented in 1859.

Cochineal: a red dyestuff derived from a type of scale insect. There are two biologically distinct types, one from the Old World and one from the New World. However, the principal component of the dyestuff is the same in each: carminic acid.

'Copper resinate': the name often given to a green glaze, usually containing verdigris. It was widely used during the Italian Renaissance due to the lack of other good greens. Often the resin was simply a constituent of the glaze paint medium, with the oil; but sometimes the verdigris was warmed with the resin (pine resin) before mixing with the oil, and the product then is 'copper resinate'. This product discolours easily to give opaque brown, or even black, with exposure to light.

Craquelure: a network of cracks over the surface of a painting which appear as a result of ageing. The cracks are caused by shrinkage, the movement of the ground, paint and varnish. These are not the same as the cracks or defects caused by the inappropriate use of oil.

Crimson: the word comes from kermes: a type of scale insect. See also *Lake*.

Dolomite: a mixed calcium and magnesium carbonate; a hard limestone, found especially in the Dolomite mountains of the Tyrol.

Earth pigments: (brown, red, yellow, orange and black) – natural silica/clay mixes, with the colour due to iron(III) oxide in various proportions. (See *Green earth* and *Ochres*)

Egg tempera: pigment mixed with egg yolk and water to make paint. The egg is the medium. (Cennini recommends that the paler yolks of town eggs should be used as the medium for painting young flesh, whereas the deeper-coloured country yolks should be used for older, rougher faces).

Emerald green: a brilliant green pigment – 'copper acetoarsenite' invented in 1814 and manufactured on a large scale from then on.

Energy dispersive X-ray microanalysis (EDX). For an outline of this process see page 14.

Fourier transform infrared spectrometry (FTIR). For an outline of this process see page 15.

Fugitive pigment: a pigment which is likely to fade in light.

Gas chromatography (GC): *eg* to find the ratio of carboxylic acids in an oil medium to identify the oil. (For an outline of this process see page 16.) Before sensitive and reliable techniques like gas chromatography, there had been much argument about the medium used in certain paintings. Some artists had made plausible copies of famous works by using egg-oil and oil-resin mixes; and, because they could copy the surface appearance of a painting, they assumed that the original artist had used the same materials and techniques as they had.

Gesso: Italian for gypsum: hydrated calcium sulfate, $CaSO_4.2H_2O$. This is the familiar white material used for ceiling coving and other plasterwork in homes and elsewhere. Gypsum, anhydrite or chalk (calcium carbonate, $CaCO_3$), mixed with glue, was used to coat panels or canvas to make a suitable surface – or *ground* – before painting began. (A partially hydrated form of calcium sulfate, $2CaSO_4.H_2O$, is Plaster of Paris, used in making plaster casts and for broken bones).

Glaze: a layer of translucent paint applied over other paint to modify its colour or to give depth and richness of colour.

Gold leaf: pure gold, beaten to a very thin sheet. It was often made from gold coins and required great skill and delicacy of handling, both in its manufacture and application.

Green earth: a marine, volcanic or igneous sediment containing silicates and iron(II) minerals, used alone or mixed with lead white. This colour was commonly used in early Italian painting for the underlayers of flesh.

Ground: this is the preparatory layer put over the support before the paint is applied.

Gypsum: a white mineral composed of calcium sulfate dihydrate; used in the preparation of *gesso*.

Haematite: a very hard red pigment, similar to red ochre: iron(III) oxide ore, Fe_2O_3.

Impasto: thickly applied paint which stands out from the surface of a picture – in relief.

Imprimitura: a thin layer applied to a basic ground to modify its colour preparatory to painting.

Indigo: a blue dye, obtained from the plant *Indigofera tinctoria* and other species, including woad.

Infrared reflectography. For an outline of this process see page 17.

Glossary

Alizarin crimson: alizarin dyestuff, the synthetic equivalent of one of the constituents of madder dye was first prepared in 1868. The lake pigment prepared from it was available by the end of the 19th century.

Anhydrite: anhydrous calcium sulfate, $CaSO_4$. See also *Gypsum*.

Azurite: a blue pigment; basic copper carbonate, $2CuCO_3.Cu(OH)_2$. See also *Malachite*.

Black: these pigments were mostly made by burning plant material such as twigs (charcoal) or peach stones. Soot was also used.

Blister-laying: sticking down paint by injecting adhesive.

Bole: a smooth clay: containing iron(III) oxide – which is usually a strong reddish-brown colour. Mixed with an animal glue or egg white it was used as the underlayer for gold leaf.

Bone black: a black pigment made from carbonised bone, which gives a warmer black than does wood charcoal.

Brazilwood: a tropical wood. The source of a cheap and very fugitive red dyestuff.

Brown earth: see *Earth pigments*.

Burnishing: polishing by rubbing. Gold leaf was burnished (usually with a smooth agate or a dog's tooth) before being punched or otherwise decorated. It was a delicate and time-consuming occupation.

Cadmium yellow: a stable dense, deep yellow pigment consisting of cadmium sulfide. Invented in 1817, available from the mid - 1840s.

Canvas: the use of canvas as a support began to dominate over wood panel in the early-to-mid 16th century. Canvas is lighter in weight than wood, it is easier to transport and cheaper to buy.

Cassel earth: a dark brown earth pigment prepared from lignite (brown coal) and containing a good deal of organic material (from peat) as well as iron (III) oxide; later known as 'Van Dyck brown'.

Cennino Cennini: an artist who wrote the earliest Italian treatise on easel painting, *Il Libro dell'Arte*, around 1390. This described traditional painting methods as practised in Italy throughout the 14th century.

Cerulean blue: a greenish blue pigment composed of cobalt stannate. Introduced in 1860.

Chloroform: trichloromethane, $CHCl_3$ – widely used as a solvent. It can also be used for killing insects, and it was a widely-used anaesthetic for operations in hospitals until about 1950.

Chrome orange: basic lead(II) chromate(VI), $PbCrO_4.Pb(OH)_2$.

Chromium oxide: (opaque) dull green anhydrous chromium (III) oxide originally used as a ceramic glaze colourant and later as an artists' pigment.

Chrome yellow: lead(II) chromate(VI), $PbCrO_4$. This pigment is rather unstable and, in an atmosphere polluted with sulfur compounds, tends to turn black. Louis Nicolas Vauquelin discovered the element chromium in 1797 (named from the Greek word for colour), and later chrome yellow. Chrome yellow is made by precipitating the pigment by mixing two solutions of ionic compounds – *eg* lead(II) nitrate and potassium chromate (VI).

Cinnabar: the mineral form of red mercuric sulfide used to make the pigment vermilion (although vermilion was also synthesised from early times).

Cleaning solvent: a great many organic solvents and solvent mixtures can be used to clean paintings and there are numerous different ones on the market.

Punching: a process of gently hammering decorative indentations into gold leaf.

Raking light: light at a very shallow oblique angle to the surface of a painting, which shows up the texture and irregularity of the paint.

Realgar: an orange-red mineral pigment; an arsenic sulfide, As_2S_2. (Mineral realgar is As_4S_4). Commonly found in conjunction with *orpiment*.

Red earth: (see *Earth Pigments*)

Red lead: Pb_3O_4 is quite rarely used in paintings, except as underpaint for the much more expensive vermilion or red lake. But examples where it has been used on its own for its bright orange do occur – *eg* in 14th century Florentine painting.

Resin: a sticky substance which oozes out of certain trees. It is insoluble in water.

Rose madder: prepared from madder root under carefully contolled conditions by a method first developed in the 19th century.

Scheele's green: a dull green copper arsenite pigment invented in 1775 but not commonly used until the late 19th century.

Scumble: a thin opaque or semi-opaque coat of paint applied to the surface of a picture to modify it. The verb 'to scumble' refers to the process of applying this layer.

Size: a gelatinous solution used to stiffen fabric or paper.

Sgraffito: Italian for scratched. A technique in which paint is applied over gold or silver leaf and then partially scraped away to reveal the shiny metal. Most often used to represent cloth-of-gold textiles.

Smalt: a blue glass, ground to powder and used as a pigment. To make smalt, cobalt ore was roasted to make blue cobalt oxide. This was heated with quartz and potash; the blue melt was then poured into cold water to form glassy particles. These were coarsely ground to make the smalt. The deepest blue smalt contains more cobalt and has the largest particle size.

Soap. For cleaning paintings. A salt of a long chain organic acid (such as those found in animal fat or vegetable oil, or a more complex aromatic acid). The soap can be designed for use under very carefully controlled conditions of pH. It must not be too alkaline, to prevent damage to the paint surface. It can then be removed from the paint surface without leaving a residue. Ordinary toilet soap contains a high proportion of sodium stearate, $C_{17}H_{35}COO\ Na$.

Sturgeon glue: a glue made from the viscera of sturgeon; superior for some uses to animal-skin glue. See also *Isinglass*.

Turpentine: an oily sticky substance which oozes out of certain coniferous trees, and is then distilled to make a volatile and pungent oil used in mixing paints and varnishes. It is a mixture of quite complicated hydrocarbons called terpenes.

Ultramarine: a bright blue pigment: a complex sodium aluminosilicate containing some sulfur. Until it was manufactured chemically in the 19th century it was obtained only from the blue mineral *lapis lazuli* mined in Afghanistan. It was more expensive than gold. *Lapis lazuli* is still mined from the same mountain in Afghanistan. 'Ultramarine' means 'from across the sea'.

Varnish: a transparent layer used for coating and preserving the surface of paintings. One common old varnish was made of mastic dissolved in turpentine. The discolouration of varnishes is the main reason why paintings are cleaned.

Verdigris: a bright green pigment; basic copper(II) ethanoate, $Cu(CH_3COO)_2.2Cu(OH)_2$. It was prepared by exposing sheets of metallic copper to vinegar vapours.

Vermilion: a bright red pigment – mercury(II) sulfide, HgS. Vermilion was made artificially from an early date, although it could also be obtained by pulverising the mineral cinnabar. It has a tendency to convert to metacinnabar, HgS, which is black.

Viridian: a green pigment – hydrated chromium(III) oxide, $Cr_2O_3. 2H_2O$. In 1797 Louis Nicolas Vauquelin discovered the element chromium (named from the Greek word for colour), and the chromium pigments viridian and chrome yellow followed later. See *Chrome yellow*.

X-ray diffraction (XRD). For an outline of this process see page 20.

X-ray radiography: lead white absorbs X-rays strongly, but so do vermilion (containing mercury), lead-tin yellow, and lead chromates. But it is the fact that lead white absorbs X-rays strongly, and is also widely used in painting, that gives an X-ray image in which the dark-light contrasts are more or less the same as those seen in the painting. Gesso does not absorb X-rays to any great extent. The total X-ray absorption in any area is the sum of the absorption in each of the paint layers in that area, so that it is possible to see whether the painting is directly on to *gesso*, or on top of blue sky or sea – both of which contain lead white in the paint mix. Flesh colours often contain lead white, so faces, hands *etc.* usually show up well in X-radiographs. Titanium white is less opaque to X-rays than lead white (**NB** Lead is used to protect hospital radiographers from X-rays).

Yellow lake: prepared the same way as red lake. The dyestuffs come from plant species such as dyers broom, weld and buckthorn.

Isinglass: a whitish semi-transparent glue-like substance, a form of gelatin made from the viscera of some fish, especially sturgeon. See *Sturgeon glue*.

Kermes: a type of scale insect yielding a red dyestuff used to make lake.

Knights of Malta: This was originally a monastic community, dedicated to Saint John. Originally known as the Knights of St John of Jerusalem (the Knights Hospitaller). The Hospital was located in Jerusalem close to the church of St John the Baptist from where they took there name. They administered a hospice-infirmary for pilgrims to the Holy Land, and later became an order of Knights when they were also required to provide military protection for crusaders.

Lac: a scale insect, the source of a red dyestuff (and also shellac – a type of varnish).

Lake: a pigment made by precipitation on to a substrate from a dye solution — *ie* causing solid particles to form which are coloured by a dye. Lakes may be red, yellow, reddish-brown or yellowish-brown. They give a translucent paint when mixed with the medium, particularly oil. They are often used as glazes and have a tendency to be fugitive. Red or crimson lakes were made originally by precipitating hydrated alumina with the dyestuffs from madder, kermes, lac, cochineal or brazilwood. Now, more usually, lakes are made with synthetic dyes.

Lapis lazuli: a complex sulfur - containing naturally occuring sodium aluminium silicate used to make the bright blue pigment ultramarine.

Laser microspectral analysis (LMA) also sometimes called **Laser microspectrography (LMR)**. For an outline of this process see page 18. A method used for elemental analysis of the inorganic materials present in paintings. This process is no longer used at the National Gallery.

Lead-tin yellow (type I): a mixed oxide of tin and lead, Pb_2SnO_4.

Lead-tin yellow (type II): $Pb(Sn,Si)O_3$ (approximately), containing some free SnO_2 and some silicon.

Lead white: basic lead(II) carbonate, $(2PbCO_3.Pb(OH)_2)$. Made synthetically from an early date. Now banned because of its toxicity.

Lemon yellow: strontium chromate(VI), $(SrCrO_4)$. Not known before the 19th century.

Linseed oil: a drying oil, made by pressing the seeds of the flax plant, which was, and is, also grown for making linen.

Madder: a red dye obtained from the roots of a trailing herbaceous plant *Rubia tinctorum*; the pigment made from the dye can fluoresce orangy-pink in ultraviolet light.

Malachite: a green pigment; a basic copper carbonate, $CuCO_3.Cu(OH)_2$ (see also *Azurite*). Both azurite and malachite occur as natural copper ores. They can also be synthesised – *eg* artificial malachite.

Mass spectrometry (MS). For an outline of this process see page 19.

Mastic: a gum or resin exuded from *Pistacia* tree species, dissolved in turpentine to make a varnish.

Medium: the glue which, when mixed with a pigment, makes paint. The two most widely used in old paintings were egg yolk (with water) – see *egg tempera* – and drying oils such as linseed or walnut oil.

Metamerism: this occurs when two colours which appear to be the same in, say, daylight no longer do so when lit by, say, tungsten lamp bulbs, or when printed in a colour photograph. It is particularly glaring with blues, and most likely to be seen because of the large areas of sky in pictures.

Naples yellow: a manufactured lead and antimony compound. It has been found in 16th and 17th century paintings but was more common in the 18th century. It varies in formula.

Ochres: yellow, brown and red – natural clay/silica mixes with varying amounts of iron(III) oxide, Fe_2O_3, as the main colouring. One type of brownish-yellow ochre is sienna; this, when heated strongly, gives burnt sienna, which is a warm red–brown. Raw and burnt umbers are similar.

Orpiment: a bright yellow mineral pigment; an arsenic sulfide, As_2S_3. Commonly found in conjunction with *realgar*.

Patination: a pleasing appearance of age normally acquired through the passage of time, but sometimes simulated by forgers.

Pendants: paintings often of identical size and complementary subject matter, painted to hang together.

Pentimento (plural - pentimenti): a visible alteration – where an artist changed his/her mind, as the work on a painting progressed. The word comes from Italian 'pentire' – to repent.

Pouncing: transferring a design by dusting coloured powder (usually crushed charcoal), through holes pricked along the outlines of a drawing on paper or parchment.

Priming: a preparatory coat put onto the support or ground to prevent subsequent layers from being absorbed.

Prussian blue: a complex iron cyanide of approximate composition $Fe_4[Fe(CN)_6]_3.H_2O_{14-16}$. First discovered accidentally in Berlin, between 1704 and 1710.

11

Technical glossary

Energy dispersive X-ray (EDX) fluorescence analysis. Whereas X-ray diffraction (XRD) separates the characteristic X-rays of a material by their wavelengths, EDX does so by their energies. The data can be displayed as peaks on a graph, where energy is on the horizontal axis and intensity is on the vertical axis. Elements are identified from their characteristic patterns of peaks. Energy dispersive X-ray apparatus is compact, and can be linked to a scanning electron microscope (SEM) – hence SEM-EDX. Extremely small samples can be studied. Also, SEM-EDX can show not only which elements are present but also their relative amounts. From this and evidence from optical microscopy, particular pigments can be identified. SEM-EDX, unlike XRD, is a primary analytical tool – and an important first-choice method of analysis, see page 20. X-ray fluorescence is used by many museums and galleries as a non-destructive technique – without a paint sample being taken. It can be used directly on the painting or object, and gives the overall pattern of elements present in the area examined. What it cannot do is pin-point where in the paint layer structure – from the surface to the ground – a particular element is.

Fourier transform infrared spectroscopy (FTIR). Every pair of atoms joined by a bond in a molecule is vibrating. The frequency of the vibration in a particular bond depends on which atoms are joined by the bond, and also to some extent on what other bond vibrations are happening nearby in the molecule. If infrared (IR) radiation is shone through a sample of a substance, the frequencies of radiation which are absorbed usually can tell us which kinds of bonds are in those molecules. A conventional IR machine gradually changes the frequency of the radiation passing through the sample – only one frequency goes through at a time. Over a period of several minutes the machine produces a plot of absorption against frequency. The peaks on the plot give information about which kinds of bond are in the molecule.

In FTIR all the IR frequencies go through the sample simultaneously. The same frequencies are absorbed as with the ordinary IR, but in FTIR a computer interprets the information carried by the radiation which passes through the sample. The mathematics involved is complicated, and is called a Fourier transformation after the early 19th century French mathematician who developed the mathematical theory. The spectrum plot looks the same as for ordinary IR. Fourier infrared spectroscopy is much faster and more sensitive than conventional IR. Also, a spectrum can be obtained from a very small (pinprick size – about 1 mm ^{2}) sample by adding together the information obtained from several scans of the sample, if the equipment is linked to a microscope. This is obviously useful when analysing tiny flakes from paintings. Fourier transform infrared spectroscopy has often been used to find which medium was used for the paint, but although it can identify an oil, it cannot identify what kind of oil – *eg* linseed, walnut, poppy – is present, because they have the same types of bonds, but arranged differently. Fourier transform infrared spectroscopy leaves the sample intact. It is non-destructive, although it could be argued that taking a sample from a painting is in itself a destructive process.

Gas chromatography (GC). You may be familiar with paper chromatography, and the basic principle here is the same. In paper chromatography, the paper is the stationary phase and water or some other liquid is the solvent or mobile phase. As the mobile phase flows past the stationary phase, the different solutes carried by the mobile phase are held back to different extents by the stationary phase. In GC, a non-reactive carrier gas is the mobile phase, and the stationary phase is a solid, or a non-volatile liquid coated on inert solid particles, held in a long, coiled narrow column in an oven. A small sample is injected into a heated chamber. The sample is swept slowly through the column by a carrier gas, such as helium or nitrogen. The column can be held at the same temperature or gradually heated. The different solutes emerge from the column into the detector at different times. The detectors used are quite complicated, and usually work through changes in thermal conductivity or by flame ionisation. Sample sizes can be very small. Gas chromatography can be particularly useful in deciding which medium has been used in making the paint, or what is in the varnish which protects the surface of the painting. The sample has to be treated chemically to release the different components in the medium, and also to make them sufficiently volatile. Once that is done, egg tempera gives a gas chromatogram which is very different from those of the drying oils. Further, it is easy to distinguish linseed oil, walnut oil and poppy seed oil by the relative areas of certain peaks in their chromatograms. If the peaks are very sharp, the peak heights can be used. Because of the chemical reactions involved and the separation of components, GC destroys the sample – it is a destructive analytical technique. In gas chromatography – mass spectrometry (GC–MS), the output from the GC is put into a mass spectrometer. This very powerful (and very expensive) method is used to detect traces of minor constituents of the paint medium that are too small to be detected by GC. It has also been used to detect traces of drugs in athletes and racehorses.

High performance liquid chromatography (HPLC). In ordinary liquid chromatography, the mobile phase (the solvent, plus solutes) runs by gravity through a vertical column which contains the stationary phase. This could be particles of silica gel, alumina, or cellulose. The different solutes move down the column at different rates. If they are coloured, bands of colour separate as they move down. In HPLC the stationary phase consists of uniform very small porous silica particles (typical diameter 10^{-6} m). This results in a very high surface area, which gives much better separation of the solutes. But small particle size means that liquid would only run very slowly through the column, so a constant high pressure (up to 100 atm) is applied, to push the liquid through. The column is normally 10–30 cm and is made of metal to resist the pressure. Once more, small sample sizes can be used. The sample is destroyed in the process. The method can be used for analysing the medium in paint, and for examining large and delicate molecules which would be destroyed in GC or by heating. In the National Gallery it has been used to study dyestuffs used in lake pigments.

Infrared (IR) reflectography. Some wavelengths of visible light are absorbed when it passes through a transparent coloured material – *eg* stained glass. What you see is what gets through. The wavelengths which are absorbed give information about the substances in the material which the light passes through. But visible light cannot get through the pigment on the surface of a painting. So, instead, light is shone on to the surface at right angles to it. The wavelength of the light is changed gradually across the spectrum. The surface of the painting is rough and reflects light in all directions. Light reflected at, say, 450 nm is collected and examined. The light which is reflected is the light which is not absorbed by the pigment. (If a paint looks green, it is because it reflects green and absorbs the other colours). A reflectance spectrum is therefore a kind of opposite to the absorption spectrum, and gives much the same information. Infrared radiation is at wavelengths which are longer, and therefore at frequencies which are lower, than visible light. Infrared reflectography is particularly good for detecting the underdrawing in a painting – *ie* where the artist has used charcoal, graphite pencil or black ink (all involving carbon) to mark out the design. Carbon absorbs IR strongly, and IR can penetrate surface layers of pigment better than visible light can, if the paint is not too thick and does not contain a pigment – *eg* carbon black or azurite – which absorbs IR. This technique is non-destructive and provides information about the whole painting at once.

Laser microspectral analysis (LMA). This is definitely a destructive analytical technique! A high-energy pulse of laser light vaporises the sample. Laser light is monochromatic – *ie* all the light has the same frequency. The energy is enough to vaporise even the crystalline compounds of metals which most of the old pigments are made of. The tiny plume of vapour from the sample rises to the gap between two electrodes. A high voltage spark between the electrodes then excites the atoms and ions in the vapour – *ie* some of their electrons are given extra energy, and move into higher energy levels. As they return to their original energy levels, light is given out – an emission spectrum is produced. The frequencies of the emitted light depend on the element or elements involved, and so can be used to identify pigments. This technique is very sensitive and can be used with pigment samples as small as 10^{-7} g; but it was never widely used and has mostly been replaced by SEM-EDX (p14).

Microscopic analysis. Before modern analytical machines were invented, this was virtually the only technique available for scientific analysis of paint layers. Skilled and experienced workers could gain a great deal of information by using it. It is still widely used. Usually, a paint sample is embedded edgewise in a resin and cut and polished to expose its layer structure. A good sample shows the gesso ground, possibly the glue which seals the ground and the material used for underdrawing, the underpaint, and the surface paint layers. An experienced operator can tell from the colour, shape and size of the particles which pigments are present, and even whether hydrated or anhydrous calcium sulfate (gypsum or anhydrite) has been used for the ground. The use of polarised light, which interacts strongly with crystals – on a dispersed sample, not a cross-section – provides further information. Inspection of particles in the top layers can show whether a pigment has faded. The electron microscope and its variations – such as the scanning electron microscope (SEM) – can reach far higher magnifications than can optical microscopes, and can give direct information about elements present.

Mass Spectrometry (MS). This technique is used for analysing organic materials such as paint, binding media, resins, and varnishes. Modern MS machines can use nanogram (10^{-9} g) or even smaller samples. The method needs samples to be at least reasonably volatile, so is usually used for the medium rather than the pigment. The sample is vaporised (and so destroyed). Some of the molecules in the vapour are made into positive ions by bombardment with a beam of electrons – if one electron is knocked out of the molecule M, it becomes the ion M^+. Because ions are charged, they will respond to an electric field. A high voltage can therefore be used to accelerate the ions, which are focussed into a thin beam. A moving ion will be deflected by a magnetic field, but the amount of deflection will depend on the mass/charge ratio of the ion. So different ions fly though the magnetic field on different curves, and arrive at a detector. A mass spectrum of the sample is produced, with the mass number – *ie* the number of protons + neutrons in the particle – on the horizontal axis, and the peak heights giving the relative number of particles with the different masses. Under these conditions, big molecules break up into smaller pieces, and this fragmentation pattern gives a great deal of information about what is in the original molecule. For all this to happen, the inside of the mass spectrometer must be kept under vacuum so that the ions do not collide with air molecules. As stated earlier, MS can be coupled with GC to give an extraordinarily sensitive method of detecting – *eg* traces of drugs in urine or blood. A complex mixture can be separated by GC and each pure component can have its molecular structure examined by MS.

X-ray diffraction (XRD). This technique is used for the characterisation and identification of crystalline pigments. The structures of solid materials can be studied using X-rays – but NOT by shining the X-rays right through (as, for example, when finding out whether your ankle is broken, or whether Titian changed his mind while painting Ariadne). Instead, use is made of the fact that the wavelength of X-rays is comparable with the distance between the layers of atoms in a regular crystal lattice. If X-rays of a particular wavelength are shone on the crystal at an angle, some X-rays are absorbed but others pass straight through the crystal. The X-ray energy absorbed by the different layers of atoms is re-emitted as X-rays: it could loosely be said that some of the X- rays are reflected by the layers. As the angle at which the X-rays hit the crystal changes, the X-ray waves reflected from the different layers alternately cancel each other out (causing a dark region on a photograph) and reinforce each other (causing a bright region). What is called a diffraction pattern is produced. You can see a diffraction pattern if you look at a small bright light through a stretched piece of fine fabric such as silk. The pattern is caused by light waves interacting with the regularly-spaced threads of the fabric, just like X-rays interacting with the layers of atoms in a crystal. You can also see diffraction effects – this time splitting up white light into the colours of the spectrum – when light shines on the surface of a CD or at a shallow angle on the surface of an old LP record. Sir William Bragg and his son Sir Lawrence found an equation which linked the wavelength of the X-rays, the angle at which they hit a crystal, and the distance between the layers of atoms in a crystal. This equation provided a basis for working out simple solid structures, and the Braggs were awarded a Nobel Prize. Big molecules were a much more difficult problem, but methods were eventually found for materials such as vitamin B_{12} (for which Dorothy Hodgkin won a Nobel Prize). However, the crystals in gesso ground or in pigments are very small indeed. Fortunately, every crystalline substance has an unique X-ray powder pattern, produced when X-rays are shone on the millions of randomly-arranged tiny crystals in the powder. These patterns are now recorded and stored internationally, and when a paint flake is examined by XRD the crystalline materials in it can easily be identified by comparing their diffraction patterns with those held in the store.

The Chemistry of Art

The paintings

The structure of paintings

Before starting to use this pack it is important to become familiar with the layer structure of paintings. The information below relates to the typical European panel or canvas painting dating between about 1250 and 1900 AD. (Wall paintings – murals – whether done in fresco or another technique are not included here.)

There are three main parts to a painting.

1) **The Support.** This is what the painting is painted on. It is usually either a specially prepared wood panel, or canvas.

   Wood. Oak panels were commonly used in northern Europe, while poplar was more usual in Italy.

   Canvas. Canvas (usually made of linen) stretched tight on a stretcher was used from around 1460 particularly in Italy. However, for about 200 years painters used either wood or canvas. Canvas gradually became the more popular support.

2) **The Ground.** This is a preparatory layer put on to the support before the paint is applied.

   For panel paintings, the ground might consist of a layer of linen glued to the wood followed by several layers of **gesso** or chalk, smoothed to give a hard white enamel-like finish. Sometimes the linen layer was omitted. If any large area of a painting was to be gilded – *ie* covered with gold leaf – this was usually done before the paint was applied.

   Canvas takes less preparation and can be primed with a glue **size** and then given a paint undercoat, although in earlier examples one or more layers of gesso on canvas are also sometimes found.

3) **The Paint.** Paint has two components: pigment (coloured matter ground to a powder) mixed with a liquid binder (or **medium**) to make it stick.

Many early panels were painted in **egg tempera**, in which the medium was egg yolk mixed with water. This technique reached its peak in 14th century Italy.

In an oil painting a drying oil (such as linseed or walnut oil) was used instead of egg. Oil had been widely used in northern Europe for many centuries. By the mid-15th century its use spread to Italy, and some paintings show both types of medium. By the early 16th century, oil had almost totally replaced egg tempera as a medium throughout Europe.

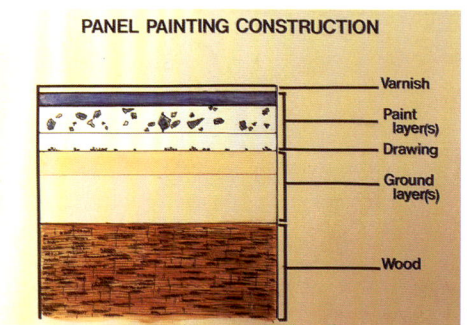

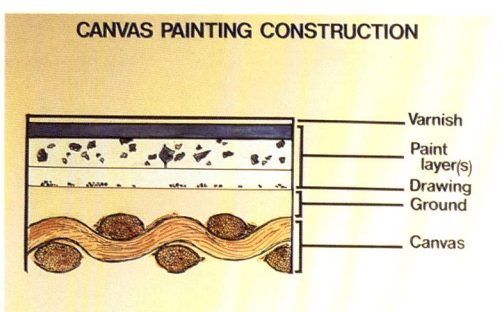

1. *Saint John the Baptist with Saint John the Evangelist (?) and Saint James*

Artist	NARDO di Cione (active 1343; died between 1365 and 1366) (pronounced Chee-oh-nay)
Medium	Egg tempera
Support	Poplar
Size	159.5 x 147.5 cm
Date	About 1365

This painting, which is over 600 years old, is a good example of the use of egg tempera paint and gilding. When it was cleaned in 1981-2, the materials used were analysed, and some of the results are discussed overleaf.

The subject

This large painting shows three standing saints, who can be identified as Saint John the Baptist in the centre with, on the left, probably Saint John the Evangelist (in a green and pink robe), and on the right Saint James (with a pilgrim's staff and book). The picture almost certainly was an altarpiece in San Giovanni della Calza (then San Giovanni Gerosolomitano) in Florence, a church which belonged to the **Knights of Malta**, and this would account for the central position of Saint John the Baptist.

The whole picture-making process involved teamwork, with many specialists (such as frame-makers and gilders) and apprentices being involved.

The support, ground and underdrawing

The picture is on poplar, the most commonly used wood for Italian panel paintings. Four samples (from different parts of the painting) examined by **gas chromatography** showed the **medium** to be egg. The picture is of high quality and it is in very good condition. The techniques displayed in this picture are extremely close to those described in **Cennino Cennini's** treatise *Il Libro dell'Arte* which was written around 1390, some 30 years after the painting was made. The poplar **support** was covered by a layer of **gesso** which was smoothed to give a uniform white enamel-like surface. This is called the **ground**. Here it consists of a single fairly thin layer of a gesso containing both **gypsum** and **anhydrite**; this was shown by **X-ray diffraction**.

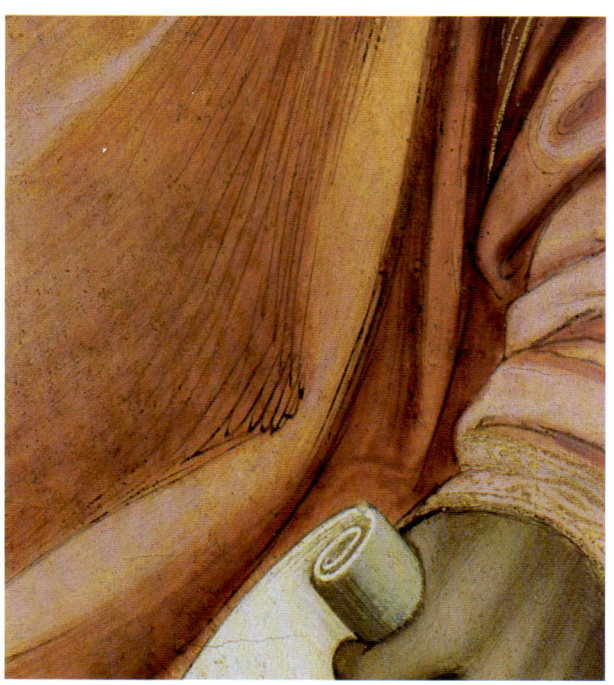

The design was then drawn on to the gesso. Because paint tends to become more transparent as it ages, quite a lot of the underdrawing is now visible, particularly in the pink robe of Saint John the Baptist. This is also partly due to the fact that the robe is painted largely in **lake pigments** which have faded. (This is because lake pigments are prepared from dyestuffs which are not stable to light). The underdrawing can be seen with the naked eye, though it shows up better when viewed by **infrared reflectography**. Examination of paint samples and infrared reflectography imaging suggest that it was done with a quill pen, using a liquid medium (*ie* black ink) rather than with charcoal.

Saint John the Baptist's pink robe showing underdrawing

Gilding

A section of the background - reddish-brown bole showing through the gold

The areas to be gilded were given a layer of **bole** (a mixture of smooth, red clay – containing iron(III) oxide – and animal glue or egg white) which was **burnished** (or polished) when it had set hard. **Gold leaf** was then applied using an aqueous adhesive such as glue **size** or egg white, and then burnished with a special tool made from agate or a dog's tooth. A measurement of the thickness of the gold on another painting of about the same date was about 256 nm (that is 0.000256 mm). Here, some of the gold in the background is worn – so the reddish bole can be seen in places.

The haloes are perfectly circular. Dividers would have been used, the central point sometimes being sited in the corner of a saint's eye to prevent detection, and the other point used to incise the circle in the gesso. The gilding of the haloes is **punched**, or indented with a pattern. It was a very skilled job to do this without piercing the gold leaf. These haloes are exceptionally fine. The dating of the picture involved detective work on the punch marks; the set of punches used here was most probably brought to Florence by an artist called Giovanni da Milano in the early 1360s.

Sgraffito

The floor covering was done using a technique known as **sgraffito**. In this painting, gold leaf was laid on bole; this was then painted over completely using **red lead** in **egg tempera**. (shown by X-ray diffraction and **laser microspectral analysis**). Then the pattern was applied. One section would have been drawn on parchment or paper, and used repeatedly to create the repeating pattern. The outline of the pattern was achieved by **pouncing** – dots of powdered charcoal or lead white were dabbed through holes pricked around the outlines of the pattern. **Ultramarine** in egg tempera was added on top of the red lead, for birds and flowers. The paint was then carefully scraped away while still soft (using a bone or wood scraper) to reveal the gold beneath. All over the resulting brocade or 'cloth of gold' the gold was then 'grained' with a rosette – a small multipoint punch.

Examination of the paint edges here show them to be softly curved, so the paint was not fully dry while it was being scraped. This is what you would expect as you cannot scrape paint for sgraffito if it has dried too much.

Once the gilding was completed the picture was painted. The paints were made in the workshop as needed.

The pigments

Scientific examination of the painting has revealed that the following pigments were used. For Saint John the Baptist's pink robes there is a very thin layer of deep crimson, with opaque pink layers above over the black underdrawing. Nardo probably started painting the deepest shadows of the robe with a crimson lake – followed by two layers of the lake mixed with **lead white**. The red dyestuff in the lake may derive from Polish **cochineal**. The blue lining to the robe is natural ultramarine.

Saint James's blue robes also have the same kind of blackish underdrawing, which can be seen using infrared reflectography. The blue pigment is natural ultramarine; the deepest shadows are of pure ultramarine glaze over ultramarine mixed with lead white. Ultramarine was at that time obtained only from the blue mineral lapis lazuli mined in Afghanistan. It was (and if of good quality still can be) more expensive than gold, and several ounces were needed for this painting, which suggests that Nardo had been given a very important commission. Ultramarine was usually supplied by the patron or was shown as a separate item in the artist's expense account.

Saint John the Evangelist wears a pale green robe of a 'shot' fabric that appears to change between light and shadow. The robe is basically green, but has yellow in the highlights and blue in the shadows. The green colour was expected to be **malachite**, but X-ray powder diffraction showed it was in fact ultramarine mixed with an artificial yellow pigment *giallolino*, now known as **lead-tin yellow** 'type II'. This combination of pigments is unusual: ultramarine was frequently mixed with red lake and white to make mauve, but it was rarely used to make green. **Azurite**, which is a greener blue, was more commonly used for this.

The scarlet lining of Saint James's cloak and the cover of the Evangelist's book show dark patches on top of the red **vermilion**. This blackening of vermilion was known since Roman times. It is more common in wall- paintings, where the vermilion is not protected by the egg or oil medium or perhaps by varnish.

Both tempera and oil paint become more transparent with age. Also many paintings have become worn and rubbed over the centuries. As a result of one or both of these things, 'white'-skinned people in early Italian paintings sometimes seem to be a rather ghostly green. This is because, for flesh areas, an underpaint of **green earth** was used. In this painting, brown and red **earth pigments** mixed with white were largely used over the green earth for the flesh colour itself.

Marks on the gilded paint surface suggest that vertical columns originally separated the figures. These have not survived.

2. *The Virgin and Child before a Firescreen*

Artist	A follower of Robert CAMPIN (1378/9–1444)
Medium	Oil
Support	Oak
Size	63.5 x 49.5 cm
Date	About 1440

The cleaning of this painting presented particular problems as two sections had been added, probably in the 19th century. However, the cleaning also revealed that alterations had been made to the original picture which subtly changed its meaning.

Robert Campin lived in the Netherlands from c1378–1444. This painting almost certainly dates from the 1440s and is attributed to a follower of Robert Campin: it is very probable that it came from his workshop.

The subject

What seems at first sight to be an ordinary mother and child in a domestic setting turns out to be the Virgin Mary nursing the baby Jesus, the firescreen doubling up as a halo behind the Virgin's head. At the top left, through the open window, is a minutely detailed townscape showing contemporary life in the Netherlands in the 1440s: someone is halfway up a ladder, another person is mending a roof, a horseman trots by.

The picture shows a curious combination of realistic detail within unrealistic space. For example, the observation of the wickerwork of the firescreen, and the handwritten book with its jewelled clasp, are quite breathtaking in their realism – yet overall there are strange anomalies. What is the Virgin sitting on? Could she actually sit in that posture, and would her robes fold and fall like that? Does the book look as though it is really resting on the cushion? And what about the perspective of the tiled floor and the stool on the left-hand side?

Additions

If you look carefully, you will see that the top edge and the right-hand side of the picture seem slightly darker than the rest. This is because these parts of the picture are 19th century additions. How we know they are not original is discussed later on. Just above the Virgin's head is an almost horizontal strip of light colour – this is all that is left of the lintel of the fireplace after the new strip of wood had been fixed on. The 'restored' right-hand side includes part of the firescreen, some of the Virgin's hair and a small part of her robe, and the elaborate cupboard and chalice. Whoever did the 19th century restoration may have known what was missing, or it might have been entirely his or her own invention. Another version of the painting, which is now lost but of which a poor reproduction was made in 1926, showed a simple cupboard with a plain bowl on top.

The painting was acquired by the National Gallery in 1910. It was dirty, the varnish had darkened, and the retouchings of damaged areas on the painting had discoloured. But the decision to restore it was postponed until quite recently because of problems concerning the 19th century additions.

X-ray image

The **X-ray** image of the painting clearly shows not only the later additions but also that a join plus one major split and some minor ones in the original oak panel had been repaired in the past. All these areas were covered with discoloured retouching. Once it had been decided to keep the 19th century additions, the problem was how to clean the painting. The usual cleaning solvent was fine for the main, older part, but would be a problem for the 19th century section as the paint here was based on a varnish medium and would dissolve if such a solvent came into contact with it. So a complicated water-based surfactant or **soap** formula was worked out by National Gallery scientific staff, and thoroughly tested before being used on the 19th century sections of the painting. Much discoloured paint along the joint and splits was carefully removed using a surgical scalpel. Some tiny flakes of the original paint were taken from the edges of the panel for examination.

10

What cleaning revealed

General cleaning, plus removal of the darkened **varnish**, immediately revealed many things which had not been seen for generations – such as the individual drops of milk on the bared breast, and the points of yellow flame visible through the weave of the firescreen. The ring and chain for hanging pots above the fire, together with the tiny strip of lintel, emerged

A section of the firescreen

from beneath restorer's overpaint above the Virgin's head, and showed that the 19th century restoration had been incorrect at this point. Most importantly, the genitals of the Infant were found to have been deliberately painted out.

The darker toned paint on the 19th century additions was presumably matched to the already discoloured original paint. It was decided that the modern restoration would be done so that the additions remained fractionally darker. This meant that the whole composition was left intact, but that a careful viewer could spot the difference between the original and later parts.

Examination of paint flakes from damaged areas showed a major change of composition during the painting. In the background of the top half of the painting a red layer (**vermilion**) was found below the visible paint. This was confirmed by X-ray and by studying cross-sections of paint samples. The red goes down to the level of the bench seat, but not behind the Virgin. This suggests that the first idea was to have the Virgin sitting grandly in front of a red or purple **cloth of honour**.

Also, a small area of blistered and pitted paint, looking like a burn mark, was found where the original oak panel and the addition meet just below the Virgin's left cuff. Perhaps the original painting had been damaged in a fire, and the restorer replaced the wood up to the original joint in the panel? If so, the top of the panel may also have been singed and had to be trimmed and replaced.

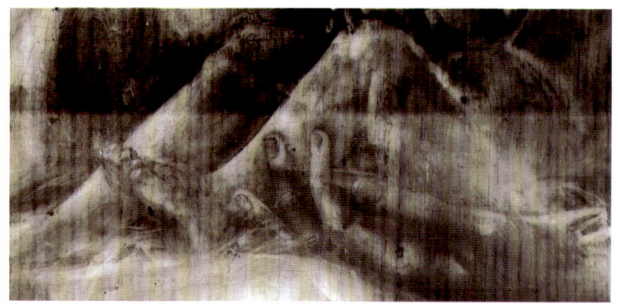

Taken from the X-ray image

The X-ray image showed up several changes of mind (or **pentimenti**) during painting. The eyes of the Child originally looked up to the bright window and his feet were in a different position. The head, hair, sleeves and cuffs of the Virgin were altered. **Infrared reflectography** revealed that the Virgin's left hand and some drapery originally covered the Child's genitals but these were later moved down by the painter to reveal them. By concealing this part of the baby at a later date, the original meaning of the painting – that Jesus was born as a normal human boy – was changed. Alterations had also been made to the Child's feet.

Examination of the support, ground and paint layers

By examining the wood grain the wood panel was shown to be oak. Oak was usually used for panel painting in northern Europe at this time. The panel was covered with a preparation layer of natural chalk bound with animal glue. This in turn was covered by a very thin brown toning layer. A staining test showed that this layer contained some protein. The underdrawing was done on this layer with a fluid containing a carbon-based pigment. Some further shading – especially in the deepest shadows of the robe's folds – was done using a dark translucent paint, while light parts – including flesh – had a thin grey underpaint. The Virgin and Child were then modelled in a second underpaint of a browner, warmer tone. The final flesh painting was done in two layers using several pigments, including **lead white**, **vermilion**, various **earths** and a red **lake**.

The distribution of media within these many layers was certainly complicated. Techniques used to identify these included **gas chromatography** linked to **mass spectrometry**, study of individual layers by **Fourier transform infrared spectrometry**, and staining tests for proteins, such as those found in egg. Results showed that the upper layers of paint were bound in **linseed oil**, while **egg tempera** was used in the underlayers for the flesh and robes as well as in the upper layers of the Virgin's flesh. Part of the sleeve in the right-hand addition was found to contain **Prussian blue**, and the golden thread in the hem of the Virgin's robe at the far right was painted with **Naples yellow**. Neither of these pigments was available in the 15th century.

A colour change due to exposure to light

The cleaning of the painting revealed some horizontal folds in the Virgin's left sleeve at the bottom right of the cuff, which had previously been hidden by restorer's overpaint. It showed that this area of sleeve was more purple than the rest of the robe: evidence that there had been significant colour fading in the drapery.

It seems that the Virgin's robe was originally a purplish-mauve. Analysis of the pigments has shown that the deep purple shadows of the robe contain **ultramarine** with red lake (probably **madder**), while the lighter areas also include lead white; the greyish-green lining of the robe contains **azurite** and lead white.

Cross-section examination of paint samples showed fading of the red lake component of the mixture, leading to two probable changes from the original colour of the robe: an overall lessening of the purplish tone of the robe due to fading of the red lake part of the colour mix, and a greater contrast between light and shade because of the greater loss of colour of the red in the lighter parts of the robe.

3. *Portrait of Alexander Mornauer*

Artist	The Master of the Mornauer Portrait (active about 1460–80)
Medium	Oil
Support	Conifer wood
Size	44 x 36 cm
Date	About 1470–80

Additions and alterations to this picture may have been a deliberate attempt to make it look like the work of a famous artist and thereby increase its value.

The subject   When this picture was purchased by the National Gallery in 1990, the sitter
was against a striking blue background and he was wearing a skull cap.
Examination showed that the blue was painted over a brownish 'wood-bark'
background and that the hat was originally taller. This raised the following
questions: *when* were these alterations made, and perhaps more interestingly
– *why?*

The picture before cleaning
Notice the blue background and the smaller hat

The picture after cleaning

Early history   The early history of this painting is not known. Nineteenth century
engravings suggest it was almost certainly in the great house at Stowe, having
been acquired by the Marquess of Buckingham at some time between 1788
and 1797. It was first recorded in the 1797 guidebook to Stowe as *Martin Luther*
by Holbein. Certainly a blue background was commonplace in portraits by
Holbein and his contemporaries in the 16th century.

However, the sitter is not Martin Luther. He can in fact be identified by the
letter he is holding as Alexander Mornauer, who became Town Clerk of
Landshut in Bavaria in 1464 and was replaced in 1488. His signet ring has the
image of a 'Moor' on it. (Moor is an old-fashioned, and to the modern way of
thinking a rather offensive, name for a Northwest African Muslim – and is of
course a pun on Mornauer's name.) The identity of the painter is not known,
but it certainly cannot be Holbein who was not yet born when Mornauer
was alive.

15

A blue background that could not be original

The blue paint of the background was examined by microchemical tests, **energy dispersive X-ray microanalysis** and **X-ray diffraction** and was found to be **Prussian blue**. This colour was not known before the early 18th century – over 200 years after the painting was originally made. So the blue background had to be a later addition.

Could this have been a deliberate attempt to raise the value of this painting and make it more appealing to an 18th century English aristocrat? A Holbein portrait of someone famous like Martin Luther would be of far more interest to a collector than an unknown artist's portrait of a mere town clerk. The transformation of the original painting certainly happened between the early 18th century – when Prussian blue first became available – and 1797, the year when it was first recorded at Stowe. It was sold from Stowe in 1848, and then sold again in 1866 – by which time it was attributed to Dürer – another big name!

After its purchase by the National Gallery in 1990 the painting was examined closely.

The panel is a soft wood, probably pine or fir. It consists of two horizontal pieces; the join goes through Mornauer's upper lip. The top edge has been cut down, possibly to remove wormy wood. The white **ground** was shown by X-ray methods to be **dolomite**, a kind of limestone found especially in the Dolomite mountains of Tyrol. So the painter may very well have been Tyrolean rather than Bavarian.

Infrared radiation and **infrared reflectography** show detailed underdrawing, which consists of two types of carbon-black pigment directly on the white ground. The paint layers are directly over the underdrawing: there is no **imprimitura**.

The cross-sections taken from the background and the X-radiograph showed that the blue paint filled cracks in the original wood-coloured underlayer, proving conclusively that this blue layer was a later addition. The blue overpaint contained **lead white** as well as Prussian blue. The **medium** for the overpaint was found by **gas chromatography** to be heat-bodied poppy oil – *ie* poppy oil partly pre-polymerised by heat – which appears not to have been used as early as the 15th century. The overpainting was done over two layers of old natural resin **varnish**, the presence of which was shown by its fluorescence in ultraviolet light. The binding medium of the original painting was shown – also using gas chromatography – to be **linseed oil**.

The blue overpainting was certainly not done because of damage to the painting, because the background has emerged from beneath it in a very well-preserved state. Cleaning showed that the painting was in excellent condition. The shadow of the hat is now clearly visible. After cleaning, the letter became not only more clearly visible but transcribable – the name Mornauer is clear. The tiny head on the seal ring on the left thumb also became clear. The textures of the fur and hair are particularly impressive.

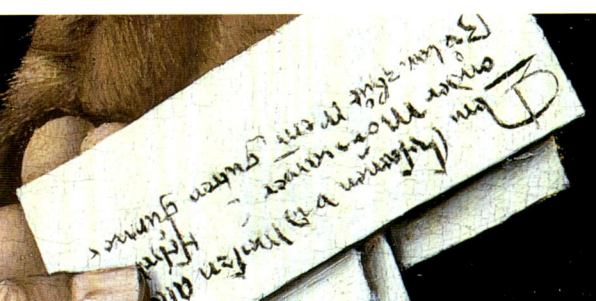

Detail of the sitter's ring   *Detail of the letter*

Pigment analysis showed that the original range of purples – some almost crimson – used in the garments has now discoloured to brownish clothes and an almost black hat. The robe was painted using blue **azurite** mixed with red **lake**. This mix is fairly translucent in oil, and – like pure azurite paints – it darkens and becomes brownish with time. This is probably due to a chemical reaction between the copper-containing blue pigment and components of the oil. The original colour of both robe and hat was a translucent purple-brown. The reddish under-robe is painted using **vermilion** and red lake.

4. *An Allegorical Figure*

Artist   Cosimo TURA (before 1431-1495)

Medium   Oil and egg

Support   Poplar

Size   116.2 x 71.1 cm

Date   Probably late 1450s

X-ray photographs of this panel painting showed that there had been substantial changes to the composition made at an early stage. Were these by Tura, or was he painting over a picture made by someone else?

The subject This is one of a series of the Nine Muses painted in a room of the Villa Belfiore in Ferrara, Italy. Tura is known to have been working in the room from 1459–63. The figure may represent Calliope (who was associated with justice). The meaning of the picture is not clear. Why is she seated on such a strange throne? Why is the bodice unlaced lower down? Does the bulge suggest pregnancy? What is going on in the cave at the bottom right? What is she holding in her right hand and why? And so on.

An **X-ray** photograph complicates things further, as it shows that originally the throne was backed by what look like organ pipes. Could this perhaps be Euterpe, the Muse of Music?

The alteration was drastic – the subsequent painting is different not only in **medium** (oil instead of **egg tempera**), but also in composition and colour, and almost certainly in subject matter.

It is one of the earliest Italian paintings in the National Gallery to be principally in oil, and the earliest to be done in a Netherlandish technique.

X-ray image. Notice what appear to be organ-pipes

19

An early Italian oil painting

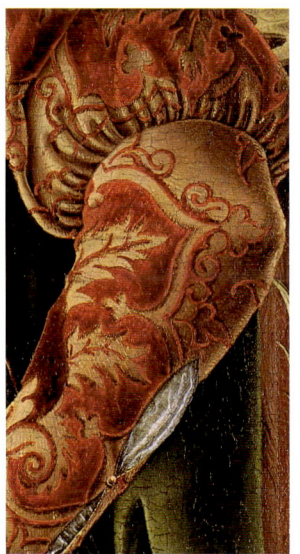

Cloth-of-gold sleeve

The painting is of great technical interest because it was done at the time when Italian painters were changing from egg tempera to oil (indeed, both media are here, with oil on top of the tempera). It is likely that Tura was influenced by paintings and/or people from northern Europe, where the oil medium had been in common use for centuries. It is difficult to believe he could have used oil paint so successfully without direct instruction. Some paintings by the 15th century Netherlandish painter Rogier van der Weyden are known to have been at Ferrara at this time. Certainly the 'cloth of gold' sleeves of the figure here are in a fabric which is very similar to those seen in Netherlandish paintings, as is the sequence of paint layers.

The panel is poplar (identified by microscopy). It was heavily flawed and cracked, and a patch of wood inserted in the plank by the original maker was pushed forward by movement of the fault and caused damage to the **ground** and paint layers. To cover faults a layer of fine **canvas** was glued over most of the panel before the **gesso** ground was applied.

The varnish had been deliberately brown-tinted. In 1866 the painting had been sent to a restorer in Milan who used the brown earth pigment **Cassel earth** in **varnishes** for artificial **patination** of pictures. Fortunately the varnish and retouchings were easily and completely removed with alcohol.

Flaking was due to poor adhesion between paint layers, especially in the green areas. In 1939 the remaining blisters and flake edges were fixed using **sturgeon glue** with an electrically heated spatula.

Cleaning intensified the **craquelure**; but, since such a surface cracking is characteristic of many old paintings and typical of those by Tura, during restoration the cracks were not restored but reduced in width to look like the cracks in the less-affected parts of the picture. Such retouching needed great precision, especially on the face.

The paint layers

The sequence of paint layers is often exceptionally complex. The evidence from the X-ray and **infrared** investigations, coupled with microscopic analysis of many paint samples to establish pigments, and **gas chromatography** to find the binding media, suggests that the painting developed like this:

The first design was drawn on the prepared panel. The main curves and lines of the columnar throne were lightly cut (incised) into the gesso. The figure was drawn with a brush using some kind of ink. The upper part of the figure was similar to what we now see; the legs and drapery were very different. Most of the underdrawing then received one or more layers of paint. The robe had a layer of **red lead**, probably intended as underpaint for the more

20

expensive **vermilion**. The bodice and sleeves were painted pale blue using an **ultramarine/lead white** mix. Landscapes were blocked in with green artificial **malachite**, and the sky with two to four layers of **indigo** with white and sometimes **azurite**. The columns of the throne were golden yellow. In the paint samples where the bottom layer of paint still adhered, the medium in those layers was shown by gas chromatography to be egg tempera.

Before work on the painting re-started, a layer of smoke and dirt formed on the surface. In the days of open fires, candles and oil lamps this need only have taken a few years.

The interval leads to the question – did Tura do the *first* design?

The lines and curves of the new throne were incised into the existing paint and gesso, then the dais, throne, shell and sea monsters drawn in. The lower part of the figure was completely redrawn on top of the first design, so that we now have a slightly 'looking-up' viewpoint.

The new painting used early Netherlandish techniques. Opaque pigment mixtures in oil were covered where appropriate with transparent and semi-transparent glazes. Walnut oil was the medium in the paint for the white marble dais – presumably walnut was chosen for areas of light colour as it 'yellows' less. **Resin** was added to the **linseed oil** in the darker glazes to add to their richness and transparency.

Summary   It now seems likely that the original unfinished design (probably by another artist) was overpainted by Tura when he took over the project for the Studiolo at Belfiore. He adopted the technique, and to some extent the style, of the Netherlandish painter Rogier van der Weyden whose works he would have seen at Ferrara.

5. *The Incredulity of Saint Thomas*

Artist	Giovanni Battista CIMA da Conegliano (about 1459/60 – about 1517/18) (pronounced Chee-ma)
Medium	Oil
Support	Synthetic panel (transferred from poplar)
Size	294 x 199.4 cm
Date	About 1502–4

Possible poor workmanship on the original panel, combined with an unfortunate immersion in the Grand Canal in Venice, led to more than 100 years of conservation problems with this painting, but also gave opportunities to examine its structure and composition in great detail.

The original commission

In 1497 the Scuola di San Tommaso dei Battuti commissioned a new, large, painting to go on their altar in the church of San Francesco in Portogruaro, 50 miles north-west of Venice. Many documents relating to this have survived, and we know that Cima was asked to do the work with the least possible expense. However, recent research has shown that he used several expensive pigments. He finished the painting in 1504 – although he threatened not to complete it unless he was paid more than the agreed amount. Surviving account books list expenses for collecting the painting from Cima's workshop in Venice and transporting it to Portogruaro – the job must have been very difficult indeed, because the picture was 3 m high and 2 m broad and painted on wood. Cima then sued for more payments, and the court case went on until 1509.

Cima lived and worked in Venice, which was central to the European pigment trade at this time, and he would have had access to the widest possible range of pigments. In this painting he used several unusual ones.

The subject

After the Resurrection, Christ appeared to the disciples and showed them his wounds. Thomas was absent and he doubted what had happened. Eight days later, Jesus reappeared and Thomas was allowed to put his finger in the wound in Christ's side – at which point he believed. The choice of subject was a natural one for the confraternity of Saint Thomas which commissioned it. This was a charitable lay association (the 'dei Battuti' indicates that they were also penitential) which ran four hospitals. Such a subject, with Jesus risen from the dead in perfect form, would have offered great comfort to people with leprosy and other such disfiguring diseases who prayed in front of it.

Long term conservation problems

The picture had conservation problems for at least 200 years. In 1981, during the restoration at the National Gallery, an inscription was discovered on one of the floor tiles at the bottom right-hand corner of the picture. This indicated

that the picture had been restored in 1745. Many crude retouchings, presumably dating from this time, suggested that the paint was already blistering and flaking before 1745. This could have been due to a fault in the preparation of the panel – *eg* the wrong amount of glue **size** applied before the **gesso ground** was put on – or because of neglect and a poor environment.

Catastrophe

In 1820 the painting was again sent for restoration, this time to the Academy of Fine Arts in Venice. Once again there was an argument over the bill; and from 1822 until 1830 the painting was stored in a room on the ground floor of the Academy. At some point during this time a tidal surge up the Grand Canal, flooded the room and knocked over the easel on which the painting stood, and the story goes that it spent several hours under water. It was back in Venice for more treatment during 1852-54.

Bought by the National Gallery

In 1863, it was seen in Portogruaro by Sir Charles Eastlake, the then Director of the National Gallery, who was on a picture-buying expedition. He thought it was in poor condition but offered £1600 (then a very large sum); the offer was refused, but this started yet another long legal dispute in Portogruaro about who actually owned the painting. In 1869, the new Director of the Gallery, Sir William Boxall, went to Portogruaro. He was not happy about the state of the picture, but made an offer of £1800, and this time it was accepted. The Gallery then had to wait for nearly a year for an export licence, during which time yet another 'restoration' was attempted. In April 1870 Boxall was shaken to learn that the licence had been granted largely because of the condition of the painting, described by Italian art experts as 'bad' and 'deplorable'. The picture arrived in London in August 1870. The old, thick, discoloured **varnish** was thinned, new varnish was put on, a recommendation was made that 'no restoration should again be attempted', and the picture at last went on display in the Gallery in November 1870.

Continuing conservation problems

In spite of the 'no restoration' recommendation, **blister-laying** was attempted seven times between 1877 and 1938. Since no cleaning was done and the 1870 varnish was not removed, success was unlikely. More attempts were made, without success, to re-fix the loose paint; eventually the whole painting was covered with special tissue paper attached with **mastic** and **turpentine**, and stored face upwards for nearly 20 years.

The painting during conservation. The old wooden panel has been removed and the paint layers are being attached to a new support.

An inspection in 1969 found that the problem was not that the paint was coming away from the gesso ground, but that the gesso – with the paint on top of it – had flaked away from the panel. In many areas the poplar panel had suffered from fungal attack or been eaten by woodworm, so that there was nothing for the gesso to attach to.

The support and ground

It was decided to transfer the painting to a new support. (Transfer is always considered only as a last resort). Remember the size of this painting is 3 m x 2 m ; the total thickness of the gesso and paint layers together is less than a millimetre – and they are brittle.

The painting covered by a single layer of facing paper

The first job was to re-cover the whole paint surface with facing layers of a special tissue paper and adhesive. The whole altarpiece was then placed face down on a temporary support. The wood of the original panel, all 5 cm thick of it, was then slowly and carefully removed by hand using chisels, gouges and – at the end – surgical scalpels. The paint layer could then be attached to a modern fibreglass/ honeycomb aluminium support.

As in all restoration work, the part played by the chemists in the Scientific Department of the Gallery was crucial. They had the opportunity to find out in detail what pigments and other materials Cima had used.

The gesso ground is, as would be expected, a gypsum/glue mix, coated with a final layer of glue size. Some of the straight lines in the painting were incised into the gesso using a sharp point. There is some blackish underdrawing, some of it just visible to the eye. The **infrared** image shows little underdrawing: this may be because an iron-gall ink was used (iron was found in a sample by using **laser microspectral analysis)** rather than a carbon-black ink which would show up better. The basic binding **medium** is **linseed oil**.

The pigments

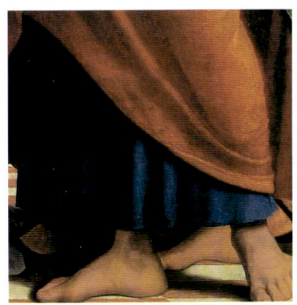

Saint Peter's under-robe is ultramarine

Analysis of paint layers, by both microscopy and laser microspectral analysis, showed that the blue in the ceiling is largely **azurite** but with an impurity which gives a greenish colour. This was probably a deliberate choice. Elsewhere in the painting, azurite is mostly used as underpaint for the very expensive **ultramarine**, but both pigments – mixed with a large amount of **lead white** – are used for the sky. The best ultramarine is used in the under-robe of Saint Peter, standing to the right of Christ as you face the picture. Lower quality pigment was used for the lesser apostle on the far left, probably to make him less prominent.

The dark red is haematite

There are three reds in this picture: red **lake** with a little **vermilion**, vermilion with a little red lake and **haematite**. For example, the robe of Saint Thomas himself is largely vermilion glazed with red lake in the shadows. Haematite has been identified in the underpaint of the upper robe of the apostle on the far left. This is a rare pigment in oil painting. One reason may be that the mineral haematite is very hard, and therefore difficult to grind up.

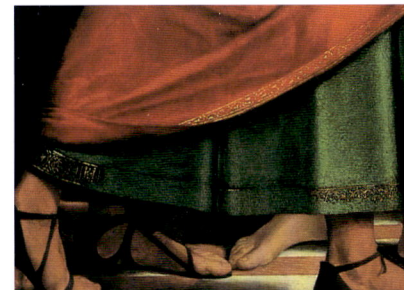

Saint Thomas's red robe is vermilion glazed with red lake

The green of the grass consists of natural **malachite** mixed with **lead-tin yellow** and some lead white. The bright green of the robes of Saint Thomas and the apostle on the far right is complicated in its layer structure, containing malachite, **verdigris** with lead white and some lead-tin yellow, glazed with **'copper resinate'** or similar.

Saint Peter's orange robe contains orpiment and realgar

The yellow of the robe of the apostle behind Saint Thomas is mostly lead-tin yellow, but with a **glaze** containing an orange-brown softwood tar. The yellow embroidery on the hem of Saint Thomas's robe and elsewhere is lead-tin yellow + yellow **ochre** and possibly yellow lake. The orange robe of Saint Peter contains the mineral arsenic sulfides, **orpiment** and **realgar**.

The black of the sandals worn by the apostles on the left probably contains **bone black** (very fine brownish particles of carbon) and vegetable black (larger, slightly shiny particles) but not the long splintery shapes of the particles of charcoal. The warm grey of the walls contains mostly bone black and lead white. The subtle optical effects – rather like a metallic sheen – in Christ's drapery are achieved using the difference between a cool grey based on the vegetable black with, over it, in some areas, the warmer, rather darker grey from the bone black.

The various flesh colours here contained lead white, red lake, vermilion, a transparent orange-brown and sometimes a little black. The paint for the body of the risen Christ also contains some pale green malachite to give it a rather deathly greyness.

Identification of most of the pigments used by Cima greatly helped the restoration. It proved possible to imitate closely the structures of the paint layers so as to achieve the correct optical effects. Also, in areas of blue and green in particular, careful choice of pigment avoided the problem of **metamerism** – *ie* two paints which appear to be the same colour in one light (*eg* daylight) but different in another light (*eg* tungsten lamp).

A happy ending!   In all, the treatment and restoration of Cima's altarpiece took 15 years. If you doubt whether it was all worth while, go to the National Gallery and up to its Central Hall. Turn left, along the great walk-way that runs in a dead straight line from the East Wing through the West Wing across the linking bridge and into the new Sainsbury Wing. *The Incredulity of Saint Thomas* is in front of you all the way, gaining in power as you get closer, the perspective in it carefully worked by Cima so that on its altar in Portogruaro, or on the wall now, the lines meet at the level of your eye, and the drama centred around Saint Thomas and Jesus appears to be happening now, in a room at the end of the vista. … And the architecture has been so designed by Robert Venturi and Denise Scott-Brown that, as you get nearer, the columns on each side of the aisle take the colour of the wall behind Jesus and gradually reduce in height to match the painting and make the illusion even more complete. Artistry and craftsmanship, old and new, combine to make a masterpiece into an intense experience.

6. Bacchus and Ariadne

Artist	TITIAN (active 1507; died 1576)
Medium	Oil
Support	Canvas
Size	175.2 x 190.5 cm
Date	1523

It was not uncommon for canvas paintings to be rolled up for ease of transport. However, this painting seems to have been rolled the wrong way – with the paint on the inside. This led to problems of flaking. Its restoration gave opportunities to study Titian's methods and materials.

The subject This large picture shows the sequel to the famous classical myth of Theseus and the Minotaur. Ariadne, the daughter of King Minos of Crete, having helped Theseus escape destruction by the bull-headed Minotaur and get through the labyrinthine corridors of her father's palace, sails for Athens with the ungrateful Theseus, who abandons her *en route* on the island of Naxos. Here she meets Bacchus, who offers himself as a husband and the sky – in which she will become a constellation – as a wedding gift. The story is known from the Latin poets (Catullus/Ovid and Philostratus).

Titian has shown the moment when Bacchus first appears. He is the scantily dressed figure with vine leaves round his head who leaps from his cheetah-drawn chariot. He is accompanied by a retinue of Bacchanalian revellers. Ariadne, who is waving hopelessly at Theseus' departing ship just visible on the horizon, turns in surprise at the sudden arrival of the god. The circle of stars above her head refers to the gift Bacchus will give her.

The patron This was originally one of a series of Bacchanalian paintings commissioned from various artists by Duke Alfonso I d'Este for a room in his castle in Ferrara, Italy.

The Duke sent Titian the **canvas** and stretcher in 1520 and for the next two and a half years badgered him to get on with it. Titian was known to be working on the picture in Venice in 1522. In January 1523 the unfinished picture left Venice for Francolino, the port of Ferrara. From there a porter carried it on his back to Ferrara (his bill for this job still exists). Titian arrived at the castle soon afterwards to finish it. It says something about the relationship between artist and patron at this point that Titian said that he would only go to Ferrara if the Duke guaranteed *in writing* his safe return to Venice!

The picture is moved to Rome In 1598 the painting was moved from Ferrara to Rome. At that time it was not unusual to take a painting off its stretcher and roll it up to make it easier to transport. However, if rolled tightly or with the paint layer inside, this can cause cracks, especially in the **gesso ground**.

When restoration started in 1967 many small paint losses and areas of disintegrating ground were found; this fits in with rolling of the canvas and rough handling, either in the porterage to Ferrara or on the journey to Rome, or maybe at some other point in the painting's history.

Conservation problems

The painting was bought in Rome and reached London in 1806 or 1807, and it was cleaned and restored around that time. However, once cleaned it was revarnished with something like **mastic** in **turpentine**; these old-type natural resin varnishes soon go yellow, and by 1826, when the picture was bought by the National Gallery, the varnish had already darkened. In 1846 it was partially cleaned, and revarnished with a mastic varnish containing some **linseed oil**. By 1894 the canvas was in a bad state, so the painting was re-lined and given a new stretcher, and again re-varnished. It was again re-lined in 1929.

The condition of the picture was examined in detail in 1967. A photograph taken with **raking light** showed the surface of the original canvas to be severely buckled and uneven. The glue used in the 1929 lining had not penetrated the back of the original canvas, so the loose ground and paint had not been re-attached. The X-radiographs also showed lots of small losses of paint and ground; most damage and retouching was in the top half of the picture, and most of it in the sky. Bacchus had badly-discoloured retouching on the right arm, head and shoulder; and pale yellow spots on his cloak were presumed to be caused by retouching with a **fugitive pigment**, probably red **lake**.

This is Bacchus and Ariadne *before the removal of the dirty varnish. The small 'clean' squares are a 'cleaning test' – small areas of old varnish have been removed. If the decision not to proceed with the cleaning had been made, the clean areas would have been made to seem dirty again!*

The 19th century **varnish** was very thick and brittle, and stuck to the paint more than the paint layers did to each other or to the ground. (The varnish was up to 80 μm thick; a single sprayed coat of modern varnish is about 10 μm. (1 μm = 10^{-6} m)). The varnish was shown by **gas chromatography** to be probably a mastic varnish with some linseed oil. Microscopical examination indicated that it was not only discoloured by age, it had been deliberately tinted brown and yellow – possibly to hide poorly matched retouchings.

The varnish was removed 100 cm^2 at a time, fixing loose paint more or less simultaneously. Laying (*ie* the fixing back on) of loose paint was done using a gelatine/water mix at 30°C and pressing, using a hand-held thermostatically-controlled electrically heated spatula.

The laying of the paint on the front had been so successful that it was decided that a new re-lining canvas, plus impregnation with adhesive of the gesso ground and paint film, need not be done. Instead, the original canvas was stuck on to a rigid, light non-warping synthetic board (with a honeycomb paper core) using a wax-resin adhesive.

The cleaning and restoration led to an enormous gain in clarity and colour (compare the 'before' and 'after' pictures). The purpose of the 1967-69 restoration was to safeguard the survival of the painting. It also gave an opportunity to study Titian's working methods and the materials he used.

Observations about Titian's colours

All but one of the paint samples were taken after varnish removal. At the time of this restoration, techniques such as **Fourier-transform infrared spectroscopy** and **energy-dispersive X-ray microanalysis** were not available, and most of the evidence from the paint samples was acquired by microscopical examination.

Titian's basic palette included **lead white**; black; and yellow, red and brown **earth pigments**. Others identified in the main colour areas were as follows.

Blue

The blue in this painting is mostly natural **ultramarine**, which is used pure in the dark parts of Ariadne's cloak and the drapery of the Bacchante with the cymbals. When it is used by itself in oil, ultramarine dries slowly. If applied thickly, the top surface shrinks and splits, and dirt gathers in these cracks and makes them show up more. Lead white mixed in with the ultramarine helps drying and stops cracking. The mauve drapery of the Bacchante with the tambourine contains ultramarine, lead white, and red **lake**. The ultramarine in this picture is of extremely high quality – intensely-coloured and pure.

But the very dark shadows may be deeper and more translucent than they originally were, and they may no longer look very blue. This is because of ultramarine's relatively low refractive index (about 1.5) and the increase in the refractive index of the oil medium with age.

Azurite was used for the blue sea and in some distant parts of the landscape. The cheaper blue pigments, **indigo** and **smalt**, are not found in this painting at all.

Green

Malachite is used here in landscape and some of the foliage. **Verdigris** and **'copper resinate'** green are also used. A major problem with copper resinate is that exposure to light makes it discolour to opaque brown or even black. Some of the foliage was painted as a single layer of green 'copper resinate' **glaze** over blue sky, and was therefore very vulnerable to light and has turned brown. An early 17th century copy of this painting shows the leaves on the extreme right still green, so the discolouration must have happened since then. However, Titian used earth pigments to paint the brown tree in the top right-hand corner, possibly as a contrasting colour, but as the green tree has turned brown, the contrast is now lost.

Red **Vermilion** was used for Ariadne's sash – a very thick layer of fine-ground vermilion, covered by a layer of larger, darker particles to intensify the colour. Crimson lake pigments are used in Bacchus' cloak and the drapery of the young faun – mixed with lead white for light and middle tones, but unmixed

and thick for the glaze in the shadows. For lakes of this period, they are remarkably thick and pure and show very little fading. Lake pigment was also used in Ariadne's flesh colour and in the pale mauve of the Bacchante with the tambourine. Layers in one paint sample suggest that the dark blue part of the drapery of the Bacchante with the cymbals was originally painted crimson or pink.

Yellow and orange

The yellow drapery beneath the urn is **lead-tin yellow**, which was shown by **X-ray diffraction** to be type I. When this painting came to London in about 1807, the painter Sir Thomas Lawrence admired this drapery, and compared the then-modern **Naples yellow** unfavourably with it. **Orpiment** (yellow) and **realgar** (orange/red) – shown to be present by both microscopy and **laser microspectral analysis** – are in the orange drapery of the cymbal Bacchante. These highly poisonous pigments are quite rare in European easel paintings, with 16th century Venice and 18th century paintings from Britain, France, Holland and the

United States providing most of the instances. High exposure to light can cause orpiment to fade to white and realgar to become more orange and then to fade. Neither seems to be significantly affected here, probably because they are more stable in oil paint than in **egg tempera**.

Titian had – and still has – an enormous reputation as a colourist. In this painting, which comes quite early in his very long career, he used in a single picture all the most colourful pigments he could lay hands on, including the uncommon ones such as realgar and malachite. And he used pigments of the finest quality; the vermilion, red lakes and ultramarine, for example, are of exceptional purity and richness of colour. He used each pigment separately in quite large well-spaced areas, and placed strong contrasting colours near to one another. He used each pigment at full strength – *ie* by itself or mixed only with white to brighten it. Sometimes the colour is intensified using glazes – *eg* the blue drapery of the cymbal Bacchante has a layer of large deep blue particles of ultramarine over a layer of finer particles. (A dark, transparent glaze over a light underlayer increases *saturation* – *ie* the apparent depth of colour – without the loss of brightness which would happen if the two layers were simply mixed.)

Ariadne's very pale skin is tinged with crimson lake pigment; the shadows of Bacchus' skin contain a green earth colour; red and brown **ochres** are used in the mixes for Bacchantes and satyrs.

The medium

The **medium** was shown using gas chromatography to be linseed oil.

Some observations about Titian's painting method

The canvas was covered by a thin layer or ground of gesso. (In this picture the gesso is a calcium sulfate/glue mix). The infrared photographs showed almost no underdrawing. X-radiographs showed fewer changes during painting than is usual with Titian, although – as discussed later – painting Ariadne obviously caused Titian problems.

Layer structure/paint cross-sections
Microscopy of paint samples coupled with X-ray photographs gave information about Titian's painting procedure and about his changes of mind during the course of painting (**pentimenti**). The layers in samples from this painting are complex, and cannot be interpreted in terms of the simple and orderly system used by earlier painters – underdrawing, underpainting, main paint layer, final glaze. Three examples are given here.

- Ariadne's scarlet scarf is painted over the flesh of her shoulder, which was painted over the blue of the sea;

- the dark blue drapery of the cymbal Bacchante was originally crimson or dark pink; this original layer was then covered with lead white followed by several layers of dark blue (crimson can still be seen at the bottom of some of the cracks in the blue paint);

- the feathery foliage – once green, now brown – was painted over blue sky; but beneath that was another sequence of green glaze (not discoloured this time, because it had been protected by the paint layers above it) over blue sky.

So the evidence shows that, unlike most early artists, Titian – at least to some extent – worked out his compositions as he went along, using little or no underdrawing.

What X-ray images can show is that lead white absorbs X-rays while gesso does so to a much smaller extent. But a very thin lead white-containing layer will not impede the passage of X-rays very much, and a very thick application of gesso or chalk may. So although lead white paint is found in the flesh paint of many thinly painted Virgins (in Italian and early Netherlandish paintings for example) these layers don't register very strongly in the X-ray because they are *thin*.

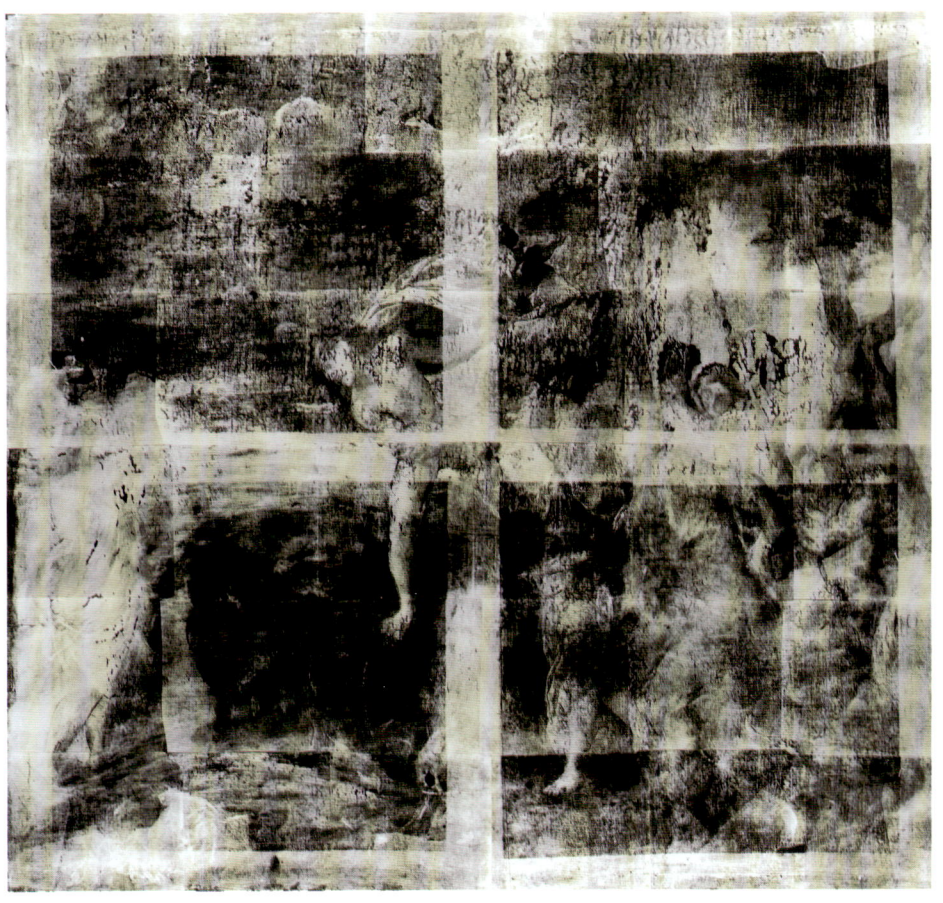

In the X-radiograph of this painting, Bacchus gives a clear, sharp image, while his cloak shows up dark. What the X-ray reveals is that the cloak was part of the artist's original design; it was painted 'over the gesso' as opposed to 'over the sky', or 'over the landscape'. The paint of the sky and distant landscape contains much lead white, and is denser and more opaque to X-rays, so shows up light in the X-radiograph. The paint of the cloak does not block the passage of X-rays and therefore shows up dark in the final X-ray image. A cross- section from this area of paint confirms this, showing only some pink underpaint (containing some lead white) between the red paint and the gesso. Also it is worth noting that the cloak is painted largely in a red lake pigment which is totally transparent to X-rays. This area of paint did not block the passage of X-rays and therefore showed up dark in the final image.

Ariadne, apart from her feet and legs, seems to have been less carefully planned. She is very dense in the X-ray because she has been altered so much.

This may be because Titian painted Bacchus early on and then had difficulty in getting Ariadne right. It also appears that the chariot, cheetahs, dog and some of the followers were painted early, direct on to gesso; but the little faun was later. (If you take a close look at the actual painting you can see the chariot wheel *through* the faun's leg).

The tree trunks were painted on gesso, but the foliage on top of blue sky. In all, the X-radiographs confirm the evidence from the paint layers.

7. *Venice: Campo San Vidal and Santa Maria della Carità ('The Stonemason's Yard')*

Artist	Giovanni Antonio Canal known as CANALETTO (1697–1768)
Medium	Oil
Support	Canvas
Size	123.8 x 162.9 cm
Date	1726–30

In this picture we find an early use of a newly synthesised colour – Prussian blue.

The subject   The view is of the open space (the campo) in front of the church of San Vidal. Work is in progress on the new façade of the church, and we see a temporary workmen's shed and various large pieces of masonry. The view is recognisable today and the well-head in the foreground still exists. From here we look across the Grand Canal to the church of Santa Maria della Carità. The campanile (or bell tower) of this church fell down in 1744. The scuola to the right of this church became the Accademia di Belle Arti in the early 19th century, and it was in this building that Cima's *Incredulity of Saint Thomas* (no.5 in this pack) was being stored when it suffered its unfortunate immersion in the waters of the Grand Canal.

Underpaint used to date the picture

The date of this picture is not known and there are no records of it before 1808. It is of uniquely high quality and may well mark a moment between Canaletto's early and mature styles. There is grey underpaint below the sky and yellow-brown under the buildings. It seems that this is what Canaletto did until about 1727–28; after that date he tended to use a uniform pale beige underpaint. So this painting probably dates from the late 1720s.

At the National Gallery

The first record is in 1808, when it was in the collection of Sir George Beaumont. It came to the National Gallery in 1828 and it was cleaned in 1852. Critics of the time said that it had been 'literally flayed', 'scoured', 'scrubbed', and 'smudged'. During cleaning, damage was noted in the right hand upper corner. No further examination or cleaning happened until 1955, when discoloured **varnish** and repaintings were removed. At that time the condition was reported as 'quite good', although some areas of sky appeared to be damaged.

By 1989 the 1955 varnish was already significantly 'discoloured'. Also, the retouchings in the sky had been made with artificial **ultramarine**. This did not match the original **Prussian blue** – the two blues displayed **metamerism**. So all colour photographs of this picture taken before 1989 show purple patches in the sky!

Some blurred clouds in the sky, near the upper right corner, were found to be painted over old paint losses, so could not be original. This overpainting could not be dated as the pigments were traditional and in use continually since Canaletto painted the picture. It is quite possible that the work was done by

Constable at Sir George Beaumont's home in 1823. Constable wrote in his *Memoirs* (for 21 November 1823): 'I have then an old picture to fill up some holes in'. It could be that Sir George wanted the painting tidied up before he gave it to the Nation. These repaints were not removed, in view of their probable historic interest, so it was decided to cover them again.

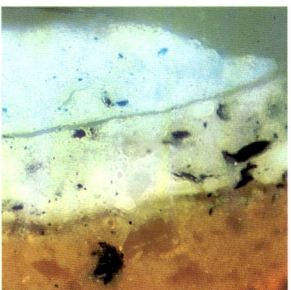

A cross-section of the sky. It shows the damage between the the pale blue top layer containing the Prussian blue and its grey underpaint. Below that is the yellow-brown layer from the upper ground.

Apart from this the picture was in quite good condition, although many small flakes were missing from the sky. Cross-sections of paint samples showed that the Prussian blue had partly come away from its grey underpaint, and the same had happened between the underpaint and the **ground**.

Cross-sections revealed a lower, orange-brown, layer of ground and an upper yellow-brown one. Cool grey underpaint was below the blue sky, but the buildings were painted direct onto the upper ground, which imparts the terracotta colour to some of them.

The sky, as usual with Canaletto, contains Prussian blue mixed with **lead white**. This is painted as a single layer over the grey underpaint, which consists of lead white and wood charcoal. Cross-sections of samples show some light-induced fading of the Prussian blue at the top of the paint layer, with less fading lower down in the layer where penetration of light diminishes. It has been suggested that the early Prussian blues – the pigment only became available in about 1710 – are liable to fading, especially when mixed with large amounts of lead white. The fading can lead to a greenish colour. The darker tints of the blue pigment (those with less white added) are more stable. A slightly unnerving feature of Prussian blue, discovered when it was used as an outdoor paint, is that some varieties can be decolourised by strong sunlight and regain their colour during the night!

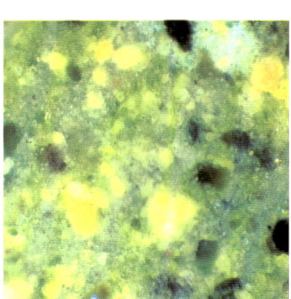

A cross-section from the green grass which is on the far side of the Grand canal in front of the church. The paint is green earth, Naples yellow and yellow ochre with a little white.

Another pigment regularly used by Canaletto is **Naples yellow**. In this painting he used it pure for the yellow jerkin of the stonemason in the centre foreground, and mixed with other colours in– *eg* the grass on the far quayside in front of Santa Maria della Carità, and in the warm terracotta of the building on the right.

8. *An Experiment on a Bird in the Air Pump*

Artist	Joseph WRIGHT of Derby (1734–97)
Medium	Oil
Support	Canvas
Size	182.9 x 243.9 cm
Date	1768

This large and dramatic painting shows an early scientific experiment in progress, and is included here more for the science in the painting than the science of making or conserving the painting.

Joseph Wright   Apart from a very few years in Italy and Bath, Joseph Wright spent all his life in Derby. He painted many portraits and classical scenes, but his most famous works are a comparative few which show scientific, technological, and philosophical themes which were of interest in the late 18th century – the time of the Industrial Revolution and of the Enlightenment – and which demonstrate his total control of the contrasting effects of light and darkness.

The subject   The air pump was invented in Oxford by Robert Boyle with Robert Hooke between 1656 – 68, and it was used by Boyle to demonstrate the characteristics of air – showing the necessity of air for combustion, respiration and the transmission of sound.

In Wright's painting, made around 100 years later, a small group of people are gathered together to watch a demonstration of the air pump to create a vacuum. The demonstrator has pumped air out of the large glass globe containing a white cockatiel which appears to be at its last gasp. The globe is a remarkable piece of glass-blowing. When the air is pumped out, it would have to withstand a very large force on it caused by atmospheric pressure. Any crack or weakness in the glass and it would implode – and glass splinters would fly everywhere.

In Wright's time, scientific demonstrations were often given by travelling professional lecturers who carried their equipment between the towns and large country houses in which they gave their shows. Wright is known to have been present at such demonstrations given by James Ferguson in Derby in about 1762. Ferguson felt that using a living animal or bird in his demonstrations was 'too shocking' (as we might also), and preferred to use a football – probably made from an inflated bladder – or artificial lungs.

The candle which provides light for the whole scene is behind the large glass vessel on the table, and its distorted reflection can be seen on the inside left wall of the vessel. The stick or straw in the vessel shows the broken appearance caused by refraction. The object in the vessel has been the subject of some debate. It has been suggested that it is a damaged skull, and that the candle represents the passing of time and the skull the inevitable result of time passing; and that the demonstrator's right index finger is deliberately pointing at these symbols of death … and that therefore, the bird is about to die.

But, as mentioned above, the demonstrator Wright knew, James Ferguson, used artificial lungs – or maybe lungs from a dead animal. When examined closely, the thing in the vessel appears to have two lobes – which is what lungs, artificial or natural, would be expected to have.

Also on the table, in front of the older girl, is a pair of small Magdeburg hemispheres. These have flat flanges, and when the two hemispheres are put together and the air they contain pumped out, it is difficult to pull them apart against the force exerted by atmospheric pressure. (The original demonstration was done many years before at Magdeburg in Germany using much larger hemispheres – two teams of horses could not separate them!)

The smaller glass vessel (in front of the little girl's right arm) contains what looks like a de-feathered goose quill – a small hollow tube, like a straw. This could be used to apply a positive pressure – *ie* to blow bubbles! – or a negative pressure so that the atmospheric pressure would push liquid up the tube and out of the vessel. Or, as we say incorrectly, to 'suck' the liquid out of the vessel.

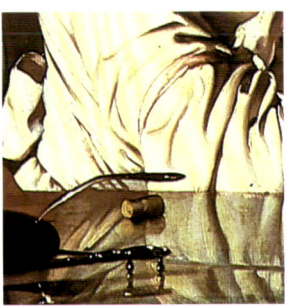

Just in front of the little girl is a cork. If you look carefully you will see that this will fit the narrow neck of the smaller glass vessel. It is probable that another part of the demonstration of what used to be called the 'science of pneumatics' consisted of lightly corking the partly-filled vessel, putting it in the globe, and then pumping air out of the globe. The cork would 'pop' out of the bottle very satisfyingly.

Opinions vary regarding the fate of the bird. In the painting, it is not yet dead. The demonstrator's left hand is on the stopcock. If he opens it in time, the bird will live; if not, it will die. The bird is a white cockatiel, an extremely rare bird to find in England at that time, and certainly not one to risk in this kind of experiment. If demonstrators did use birds, they were the common ones such as thrushes, blackbirds or sparrows. It seems likely that Wright included it for

Mr and Mrs Thomas Coltman by Joseph Wright. 1770-72

dramatic effect – to create a moment of tension, which is increased by the demonstrator looking straight at us as if asking for us to make the life or death decision; and also because its white plumage shows up so well. The actual outcome is possibly being anticipated by the boy on the right – is he lowering the birdcage ready to receive its occupant again?

Everyone is involved in the drama, except the young couple on the left who are obviously in love and oblivious to what is going on. (Almost certainly they are Thomas Coltman and Mary Barlow, who were painted again by Wright shortly after their marriage – this double portrait is also in the National Gallery).

The man in the foreground has his watch out, timing the bird's agony, the very model of the detached observer. The older man on the right sits holding his spectacles in his hand, brooding; perhaps he is thinking about death and how soon it might come, not only for the bird? Another man, probably their father, is trying to reassure the two girls who cling to each other in their distress. The boy on the left is completely fascinated.

Wright and his friends

Wright's accuracy in painting scientific equipment and technical processes is easily explained. He knew, and painted portraits of, many of the great scientific thinkers and industrial innovators of his time. Among them were Josiah Wedgwood, the pottery manufacturer; Erasmus Darwin (the very knowledgeable grandfather of Charles Darwin), who was Wright's doctor; Sir Richard Arkwright, who revolutionised the textile industry; and the mechanic and geologist John Whitehurst. These, with others such as the great engineers Matthew Boulton and James Watt, founded the Lunar Society in 1764-65. This society, through its discussions of both theory and practice, played a major part in the Industrial Revolution. A likely symbol of, or reference to, the Lunar Society is the full moon seen through the window. Members of the society met for scientific and technical discussion at each other's houses on the night of the full moon, so that afterwards – many years before street lighting – they could more easily find their way home.

Wright painted two other major scientific pictures at around this time: *A Philosopher giving a lecture on the Orrery* (1764–6) and *The Alchemist in Search of the Philosopher's Stone discovers Phosphorus* (1771, with some alterations in 1795), which is illustrated here.

41

9. Winter Landscape

Artist	Caspar David FRIEDRICH (1774–1840)
Medium	Oil
Support	Canvas
Size	32.5 x 45 cm
Date	Probably 1811

Scientific examination of this painting and its apparent 'twin' in Dortmund has given clear indications as to which one is the original. The very special atmospheric effects are largely due to the artist's understanding of the special properties of smalt.

Two similar paintings

There is a picture very similar to this in a museum in Dortmund, Germany, which until recently was thought to be the original by Friedrich. However, questions about its authenticity arose when another version turned up in a private collection in Paris in the 1980s. The picture (one or other version) had been known since it was first made, as two visitors to Friedrich's studio in 1811 wrote to each other about seeing it there – their letters still exist – and it was

The version of Winter Landscape which is in Dortmund

The version of Winter Landscape which is in the National Gallery, London

first exhibited later in 1811. It was bought in 1813 and exhibited again in 1814, but that was the last that was heard of it for well over 100 years. This second version was bought by the National Gallery in 1987.

The question to be answered was this: Which of the two versions was Friedrich's original? Might they both be by him, or was one a copy by another artist?

The subject

The picture is aptly named *Winter Landscape*. On a snow-covered hillside, a clump of fir trees and a few boulders stand silhouetted against a hazy winter sky; while the distant shadowy form of a ghostly Gothic church looms like a vision through the freezing fog, its spiky pinnacles echoing the points of the firs. Two crutches, cast upon the snow, lead us to the small figure of a man who leans against one of the rocks (a symbol of Christ, the Church, or of faith generally '…upon this rock I will build my Church', *Matthew 16:18*) and he prays before a shining crucifix half hidden among the trees. The progression of the trees leads us further along the hidden snowcovered path to the gate leading to the church precinct, and finally to the church itself. There are clues all over this picture to tell us that this is no simple winter view, but a picture with a meaning and message: that just as spring follows winter, people who follow the Christian faith will find salvation. And indeed spring is heralded by the small clumps of grass which are beginning to push their way through the foreground snow.

*Winter Landscape by Caspar David Friedrich.
Staatliche Museum, Schwerin*

This picture is one of a pair, and its meaning becomes even clearer when they are viewed together. The other, also called *Winter Landscape* (in the Staatliches Museum, Schwerin, Germany) contains two twisted dead oak trees and many tree stumps which are reminiscent of tombstones in a churchyard. The small lone figure of a man hobbles among them on crutches, and views helplessly the endless snowy wasteland ahead. This is a scene which symbolises the despair of the faithless in the face of death and contrasts dramatically with its **pendant**, the painting we are now discussing, which symbolises the hope of the faithful.

The painting now at Schwerin was rediscovered in 1941, as was the Dortmund version of *Winter Landscape*.

The Dortmund and London versions of *Winter Landscape* seem at first sight to be very similar, but in the Dortmund version there are no blades of grass pushing through the snow; no gate to the cathedral; and the cathedral itself is indistinct and shows almost no detail, unlike the London version.

Underdrawing examined

According to those who knew him, Friedrich used fine underdrawings, done first with chalk and pencil and then in detail with quill pen and ink. This underdrawing is visible to the naked eye in many of his paintings and often is important in the way the picture looks. Black ink on a light **ground**, which is what Friedrich used, shows up very clearly in infrared photos and reflectograms.

Infrared reflectograms of the London picture show some underdrawing for the rocks and trees, but show also that the underdrawing of the cathedral is *full* of minute architectural detail. There are two layers of drawing. The basic structure of the cathedral was first set out with ruled lines – possibly in pencil; thicker darker ink lines are found on top of these.

Such pen-and-ink underdrawing has been found in many Friedrich paintings, as have the short, hatched strokes used here for the trees and the stippling used for the misty sky. However, in the Dortmund version of the picture, no underdrawing shows up in infrared light; the paint surface is more even, and mostly seems to have been put on with less care than is usual for Friedrich early in his career.

This evidence combined with the fact that only *one* version was recorded in the old exhibition catalogues and by Friedrich's visitors makes it virtually certain that the London picture is by Friedrich. The other may be by him, but it seems unlikely.

Cleaning

When the painting was examined in 1987, the **varnish** had become very yellow; this affected the whole colour composition of the painting.

During cleaning the sun was discovered, to the left of the cathedral; it had been over-painted. Although it was visible in the **X-ray** and infrared images, it was not until the old retouching paint was removed that anyone could be sure that the sun was intended by Friedrich to be part of the original picture.

The ground and pigments

The ground is in two layers, but both contain chalk and **lead white** with some brown **earth** colour, with more lead white in the upper layer. Both the ground and the paint use walnut oil as the **medium**.

Restoration was mostly straightforward, although there was difficulty in matching parts of the sky because of the translucency of the **smalt** pigment which Friedrich used, coupled with his stippling technique. Only a few pigments, in a single thin layer, are used for the painting, which depends for its effect on variations in tone and texture rather than on colour.

The shades of white, grey and pink use smalt in a range of grades from pale grey to deep blue. The smalt is used alone or mixed with lead white. The pale mauve in the sky contains greyish smalt, lead white, and a little red **ochre** (mostly **haematite**). Paint cross-sections show that the blades of grass are painted on top of the snow layer; the green and brown grass contains a varying and complicated mixture of smalt, **Naples yellow**, **bone black**, ochre and possibly **Prussian blue**.

Since other good blue pigments were known, Friedrich's use of smalt may seem rather odd. But smalt has a low refractive index, and in an oil medium looks translucent. The same translucent effect cannot be achieved with **cobalt blue** (then a new pigment, which has a higher refractive index than smalt) or Prussian blue. Prussian blue in particular can be very overpowering because of its tinting strength. Friedrich used smalt for this effect in several paintings with religious or mystical subjects. A stippled brushstroke enhances the transparency and light scattering at the paint surface, and here it creates the misty, shimmering distance.

10. *Boating on the Seine*

Artist	Pierre-Auguste RENOIR (1841–1919)
Medium	Oil
Support	Canvas
Size	71 x 92 cm
Date	About 1879–80

This artist had access to a wide range of new colours made by chemists in the 19th century.

The scene is traditionally supposed to be on the Seine at Asnières, but is more probably at Chatou. The painting was acquired by the National Gallery in 1982. The **priming** is mostly **lead white**.

Pigments

The pigments used by Renoir in this painting were identified from paint samples by using optical microscopy, **laser microspectral analysis** and **X-ray diffraction**. Other than lead white, they are: **cobalt blue**; **viridian**; **chrome yellow**; **lemon yellow**; **chrome orange**; **vermilion** and a crimson coloured **lake** pigment. No black pigment was found, nor any **earth** colour. Interestingly, in his biography of Pierre-Auguste Renoir, the great film director Jean Renoir states that he never saw his father use chrome yellow. It does seem to be missing from Renoir's paintings after this date, replaced by **Naples yellow**. Jean also writes that his father was suspicious of 'newly-introduced materials'. This seems unlikely; of the pigments in the list above, only vermilion and the lake pigment would have been known to painters in previous centuries, although most would have been available before Renoir started painting.

Ready-made new colours in tubes

A pig's bladder used for storing paint

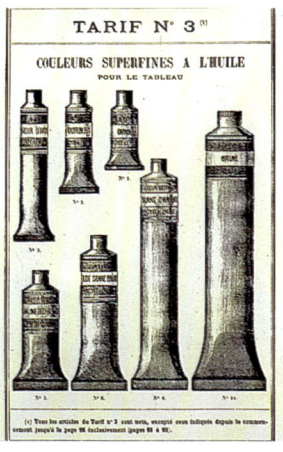

19th century paint tubes

Two new factors gave 19th century painters greater freedom than their predecessors had ever had, and encouraged the Impressionist revolution. One was the production of paint in squeezable metal tubes, with the pigment already mixed with the oil medium. Although, in the past, painters had taken ready-made paints out of doors in bladders or glass cylinders, these new paints came in tubes, could be squeezed directly on to the palette, and probably had a longer 'shelf life'. They were also actively marketed, and became attractive to amateur painters as well as professionals. The other factor resulted from the work of inorganic chemists, in experiments subsidised by the French government, who made a whole range of completely new, brilliantly coloured, stable pigments based on chromium, cadmium, cobalt, zinc, copper – and even arsenic!

Renoir exploited these new colours to the full, and in this painting he puts cobalt blue alongside chrome yellow and chrome orange to achieve something which probably no painter had ever managed before – the dazzle of sunshine reflecting off water.

Colour theories

According to the colour theory of Michel Eugène Chevreul, as expressed in his 'colour wheel' (1839), orange and blue are complementary colours; so, when they are placed together, the intensity and hue of each is enhanced. We see Renoir using this idea here. He also uses another idea of Chevreul's: by putting pink, *ie* a tint of red (the passenger's dress), next to orange – which is near red on Chevreul's wheel – he emphasises the contrast between the two colours.

How the paint was applied

The **X-ray** image of the painting shows that there were minor alterations to the prow of the boat, the reeds in the foreground, the oar and the passenger – who may originally have been lying back in the boat. The **canvas** is scarcely 'wetted' by the paint; indeed, in places the weave of the almost bare canvas shows through the paint layer, with its darker colouring having some effect on the overall appearance of the picture. The **ground**, which is lead white in oil, is thin and uneven. After the ground, next to be painted were the high-tone blues and the reflections of the women. The painting is quite fluid, and was done 'wet in wet' – *ie* an upper layer put on before the lower layer is dry. The reflection of the villa is dragged over the dry underpaint of the river. Some of the final highlights are not carefully positioned, but simply *flicked* on to the painting. For the boat and those in it, the river and the reflections, paint is applied thickly, often with brushstrokes piled on top of each other.

The surface brilliance of the painting is achieved by using opaque, almost unmixed colours. Although the range of pigments used is limited (see above), there is very little intermixing of them in most of the picture. Much of the paint has been put on 'direct from the tube' (via palette and brush!). The effect depends on unblended bright colours close together rather than on physical mixing.

The river is pure cobalt blue, with added white for lighter tones. The translucent crimson lake pigment gives a purplish undercolour on the right of the picture, but mostly the underpaints are pale yellow-green. Brilliant chromate colours are used for the boat and its reflection. Chromium pigments are also found in the greens and yellow-greens of the foreground reeds and of the river bank – the reed bank is pure viridian with some chrome yellow and lead white. The distance has the same pigments, less pure and with cobalt blue or viridian in the shadows – no black pigment at all is used. The garden walls and the sail are thick lead white.

The Chemistry of Art

Experimental section

Introduction   The practical work in this section shows students how to make their own pigments and paints. The experiments are suitable for pre-16 students working under supervision. Once the students have made their materials they can use them to create a painting of their own.

It is assumed that the Science, Art and Technology departments in your school will be able to supply materials and equipment.

Only relatively small amounts of chemicals are needed.
(See Section *Making pigments* p3.)

The experiments have been developed so that they can be done in a minimum of time. However, careful planning may be required and, at times, patience.

Any local codes of practice on health and safety should be observed, and your employer's risk assessments should be checked and followed, although we believe the activities are compatible with commonly adopted model risk assessments for science. Non-scientists may need some in-house training in the hazards of using certain chemicals.

Making pigments

Teacher's notes Pigments are the coloured materials which are mixed with a suitable medium to make paint.

A small selection of pigments can be made by students, which should provide a reasonable range of colours. You may wish to supplement those listed here with others available in your school or bought from an artists' materials shop.

Apparatus and equipment

- Fume cupboard
- Bunsen burner, tripod, gauze, tongs
- Buchner flask and filter; water pump
- Good pestle and mortar
- General glassware, beakers, flasks, test-tubes, measuring cylinders, as appropriate
- Test-tube brushes
- Sample tubes and stoppers
- Labels
- Filter papers
- Spatulas
- Glass rods
- Paper towels
- Disposable plastic gloves
- Means of drying – *eg* a radiator grille in the bench or an infrared lamp (at a safe distance above the filter paper)
- *If glass beads are to be made (see Smalt) the following are required.*
  - Lengths of nichrome wire (about 15 cm) in corks
  - Pliers.

Chemicals

- Dilead(II) lead(IV) oxide ('red lead'). (**Hazard**: toxic, may harm unborn child)
- Charcoal powder, fine
- Copper(II) nitrate. (**Hazard**: harmful)
- Copper(II) sulfate. (**Hazard**: Harmful)
- Sodium carbonate

- Lead(II) nitrate. (**Hazard**: toxic)

- Sodium hydrogencarbonate

- Calcium carbonate, fine powder

- Iron(II) sulfate. (**Hazard**: harmful)

- Iron(III) chloride. (**Hazard**: irritant)

- Potassium hexacyanoferrate(II) (potassium ferrocyanide)

- Potassium or sodium chromate(VI). (**Hazard**: very toxic. May cause cancer by inhalation. Younger students, under 13, should not handle this solid)

- *If glass beads are to be made the following are required:*

  - Cobalt(II) chloride or cobalt(II) nitrate. (**Hazard**: harmful)

  - Disodium tetraborate (borax)

  - Sodium or potassium silicate, concentrated aqueous ('water glass'). (**Hazard**: harmful, irritant, risk of serious damage to eyes)

  - Calcium hydroxide powder ('slaked lime')

Solutions

- Lead(II) nitrate (**Hazard**: toxic)   0.5 mol dm^{-3}

- Sodium carbonate   1 mol dm^{-3}

- Copper(II) sulfate   0.5 mol dm^{-3}

- Potassium or sodium chromate(VI) (**Hazard**: irritant)   0.25 mol dm^{-3}

Safety   **Safety is very important. In this section and in all the later sections, proper safety clothing and eye protection must be used.**

Disposable gloves should be used when your employer's risk assessment requires it. Their use depends on the nature of the hazard. Disposable gloves are expensive and may be slippery, increasing the accident rate. There is a need to assess the risk in using them.

Students must be supervised by staff who understand the hazards involved. Please check your LEA and/or school policies before doing any experiments.

Student's instructions

All the pigments can be stored in labelled sample tubes. Amounts need not be large; the quantities in the instructions can be reduced proportionately if necessary.

Pigments from precipitation reactions

Many pigments can be made by precipitation. (Pigments must be insoluble in the medium with which they have to be mixed to make paints, so precipitation is a logical way of making them.)

The principle of ionic precipitation works in this way.

All the common compounds which contain metals consist of ions. Ions are atoms or groups of atoms which have an electric charge. Metal ions have a positive charge; non-metallic ions have a negative charge. Examples are sodium chloride (ordinary salt), which consists of positive sodium ions and negative chloride ions; and lead(II) nitrate, with positive lead ions and negative nitrate ions.

When an ionic substance is dissolved in water, the positive and negative ions can separate from one another and can react on their own.

If solutions of two soluble ionic substances are mixed, there may be two ions in the mixture which can join together to form an insoluble substance. This causes a compound containing these two ions to precipitate. For example:

sodium chloride solution +   sodium nitrate solution +
lead(II) nitrate solution   ⟶   solid lead(II) chloride

The lead(II) chloride precipitates, and can be filtered off.

Pigments that can be made by this method include: -

Pigment	Colour
Lead white	White
Synthetic malachite	Green
Green verditer	Green
Blue verditer	Blue
Chrome yellow	Yellow
Prussian blue	Blue

For all the pigments made by precipitation, first produce the material using the specific methods shown, and then follow the procedure given here to obtain the dry pigment.

5

Procedure for making dry pigment

First carry out the precipitation reaction.

After the precipitation reaction, leave the mixture to 'mature' for a few hours before filtering. Even if using a filter pump, filtering may be extremely slow, especially for Prussian blue. It may be sensible to use an ordinary, fairly large, filter funnel and filter paper and leave it overnight to filter. Wash the residue well with deionised water. The filter paper (with residue) can eventually be opened up and placed over a radiator grille or heated gently under an infra red lamp to remove all the water. Make sure that the paper does not over-heat and catch fire. Patience may be needed!

When the solid is completely dry, carefully scrape it off the filter paper (over a large piece of paper to catch particles), grind to a fine powder with a pestle and mortar, and store in a labelled sample tube.

Most of these methods have been selected from several sources. Experimenting with concentrations, temperatures and 'spectator ions' is worthwhile. (Spectator ions are ions that are present in a reaction but take no part in it).

Lead white. Basic lead carbonate ($2PbCO_3.Pb(OH)_2$). (**Hazard**: lead salts are toxic and may harm the unborn child.)

Mix 100 cm^3 of aqueous lead(II) nitrate (0.5 mol dm^{-3}) with 150 cm^3 of aqueous sodium carbonate (1 mol dm^{-3}) at room temperature. Stir. Then follow the procedure above. **NB** You will need more lead white than any other pigment.

Synthetic malachite. Basic copper(II) carbonate ($CuCO_3.Cu(OH)_2$). (**Hazard**: copper salts are harmful.)

To 150 cm^3 of aqueous copper(II) sulfate (0.5 mol dm^{-3}) in a 600 cm^3 beaker at room temperature add solid sodium hydrogencarbonate a little at a time until fizzing stops and there is no further reaction. Stir. Then use the procedure above.

Verditers. These are also variations on basic copper carbonate. (**Hazard**: copper salts are harmful.)

The essential reaction involves adding powdered chalk (calcium carbonate) to aqueous copper(II) nitrate. The colour obtained seems to be mainly determined by three factors: the concentration of the copper(II) nitrate solution, the temperature, and the frequency and vigour of stirring.

Green verditer. Dissolve 10 g of hydrated copper(II) nitrate ($Cu(NO_3)_2.3H_2O$) in 600 cm^3 of distilled water at room temperature. Add 3.5 g of powdered calcium carbonate. Leave for at least 24 hours, stirring briefly and gently two or three times. Wash the precipitate at least three times with deionised water by decanting before filtering. Then use the procedure on page 6.

Blue verditer. Dissolve 10 g of hydrated copper(II) nitrate in 150 cm^3 of deionised water at about 4°C – *ie* cool the water in an ordinary refrigerator or use pieces of ice to adjust the temperature. Add 4 g of powdered calcium carbonate. Replace in the fridge. Stir vigorously at 20–30 minute intervals for at least six hours – *ie* during the course of a school day. Leave for at least another two days, stirring gently a few more times. Wash thoroughly, as for green verditer. Then use the procedure on page 6. Do not be disappointed if there is no great difference between the two. The conditions appear to be quite critical.

Chrome yellow. Lead(II) chromate(VI) ($PbCrO_4$). (**Hazard**: lead salts are toxic. Chromate(VI) compounds are very toxic and may cause cancer by inhalation. Disposable gloves should be worn. This preparation is NOT suitable for younger students.)

The colour depends on several factors, including reagent concentrations, temperature, and particularly pH. (To illustrate the last point, add one or two drops of dilute sulfuric acid to a little aqueous potassium or sodium chromate(VI) in a test-tube. The yellow chromate(VI) solution changes colour to the orange dichromate(VI). Adding dilute alkali can reverse the colour change). A little of this pigment goes a long way! Chrome yellow is used for painting the familiar yellow lines in streets. Mix 100 cm^3 of aqueous potassium or sodium chromate(VI) (0.25 mol dm^{-3}) with 50 cm^3 of aqueous lead(II) nitrate or lead(II) ethanoate (0.5 mol dm^{-3}). Then use the procedure on page 6.

Prussian blue. Hydrated iron(III) hexacyanoferrate(II) ($Fe_4[Fe(CN)_6]_3.xH_2O$). (**Hazard**: iron(III) chloride solution is an irritant.)

Dissolve 10.5 g of potassium hexacyanoferrate(II) in 100 cm^3 of deionised water and, separately, 13.5 g of hydrated iron(III) chloride ($FeCl_3.6H_2O$) in 100 cm^3 of deionised water. Heat the two solutions (do not boil) and mix while hot. Stir. Allow to settle and mature overnight. It is probably best to allow it to filter by gravity over a weekend or even longer – it is notorious for slow filtration, whatever method is used. Then use the procedure on page 6.

Warning – This pigment is very intensely coloured and tends to get everywhere. Dilute sodium hydroxide solution (2 mol dm^{-3}) (**Hazard**: corrosive) can be used to clean glassware and equipment. Prussian blue loses its colour with the alkali and leaves a brown stain of iron(III) hydroxide. This can then be removed with 2 mol dm^{-3} dilute hydrochloric acid solution. (**Hazard**: irritant)

(Although it contains cyanide, Prussian blue is not very toxic, which was one of its selling points when it was introduced around 1710. It used to be that in laboratories where cyanide was used, bottles of iron(II) sulfate and dilute ammonia solutions were kept ready. If anyone accidentally ingested cyanide, the contents of the bottles were poured into the victim as quickly as possible. They react with and immobilise the cyanide by forming the hexacyanoferrate(II) complex inside the body!)

A pigment made by decomposition

Iron(III) oxide. Fe$_2$O$_3$; *Jeweller's rouge* made by heating.

Iron oxide from stock jars tends to be dark brown. If this is not acceptable, a better red-brown can be obtained by heating hydrated iron(II) sulfate (green vitriol). This reaction was used in ancient times to make both oil of vitriol (concentrated sulfuric acid) and rouge – which was used both as a cosmetic and as a polish – *eg* for jewellery.

Heat about 20 g of hydrated iron(II) sulfate strongly in an evaporating basin in a fume cupboard. (**Hazard**: corrosive fumes are given off.) As the water is driven off, the solid 'cakes' and goes hard. It needs to be broken up and stirred with a glass rod. Hold the basin steady with tongs.

When all the solid is brown, allow it to cool. Grind with a pestle and mortar, and then re-heat the powder strongly, with stirring at first. Heat for about 45 minutes overall. The deep brown powder becomes a fine red-brown on cooling.

Smalt from glass beads

Smalt has been known for many centuries, and is essentially ground-up cobalt glass. The glorious blue of medieval stained glass in cathedrals was achieved using cobalt compounds.

Smalt can be made using a modification of the 'borax bead test', which was once used as a method for identifying several transition metals, including cobalt.

Push one end of a length of nichrome heat-resistant wire about (12-15 cm) into a cork. Hold the cork while heating the other end of the wire very strongly in a roaring Bunsen flame. **Eye protection is essential**.

Place the end of the hot wire in powdered 'borax' (hydrated disodium tetraborate). Some of the borax sticks to the wire, which is then reheated. The borax swells, then eventually melts to a bead. Repeat this process, as required, to obtain a larger bead. (**Hazard**: very hot material may drop off the wire.)

Dip the molten bead from time to time, as it grows, into 'water glass' (syrupy concentrated aqueous sodium silicate) and powdered calcium hydroxide. (**Hazard**: irritant and corrosive.) These materials, added to borax, make a better and harder glass.

When the molten bead is reasonably sized (5 mm or more), dip it carefully into powdered cobalt(II) chloride or cobalt(II) nitrate. Only a very small amount of the cobalt compound should be taken. Continue heating strongly until the blue colour is uniform throughout the bead. This can only be seen clearly by allowing the bead to cool and solidify. It may be necessary to take another 'dip' of the cobalt compound. Several good beads are needed to obtain a sample of smalt.

Use a pair of pliers to break the beads off the wires. (**Hazard**: pieces of glass can be projected considerable distances.) Remove the beads behind a suitable safety screen.

Pulverise the material from the beads and coarsely grind it in a pestle and mortar. Wash well to remove unreacted soluble material, dry, and re-grind. The resultant material should be grey blue. Over-grinding gives smaller particles with a paler colour*.

Pigments from stock   *Red lead.* Dilead(II) lead(IV) oxide (Pb_3O_4) and powdered charcoal can be used straight from the stock container. (**Hazard**: lead compounds are toxic and may harm the unborn child.)

*****Teacher's note:*** *Making glass beads in this way is an activity enjoyed by junior students, and can be used to reinforce points about the structure of the Bunsen flame, the nature of chemical reactions and the chemical elements associated with colour. Compounds of copper, iron, chromium and manganese, as well as cobalt, can be used to make beads of different colours. The colours obtained may vary, though not for cobalt, depending on whether heating is done in the oxidising or reducing part of the Bunsen flame. Any blue cobalt glass beads can be harvested and used for making smalt.*

Making egg tempera paint

Teacher's notes

Apparatus and equipment

- Pestle and mortar (or muller and glass slab)
- Small containers for paints – *eg* plastic pots, beakers, or film containers
- Small beakers
- Palette, or other surface on which paints of different colours can be mixed
- Flexible palette knife, spatulas
- Small plastic or wooden spoons (the type used for ice cream) are useful for transferring paint from the mortar to the container
- Disposable plastic gloves
- Paper kitchen towels

Materials

- Pigments from *Part A*
- Fresh eggs (only a few are likely to be needed). (**Hazard**: salmonella *etc*)

Cleaning

All surfaces and equipment should be thoroughly wiped with disinfectant using paper towels and then washed well with plenty of detergent. Any spills or waste should be dealt with at once. Egg tempera paints harden quite quickly and are then much more difficult to remove. Wash hands thoroughly.

Safety

See *Part A – Making pigments*

Care and good hygiene need to be exercised when handling raw eggs.

Student's instructions

In this section you follow quite closely the methods described in 14th century textbooks for artists. You should be able to mix sufficient different coloured paints with the pigments you have made.

Egg tempera paints should only be made when they are needed. They can be stored briefly in closed plastic pots or similar containers. Egg tempera paints were used to produce the older paintings described in this pack.

The egg medium. Crack a fresh egg gently on the edge of a suitable dish and allow the white to run out through your fingers (use disposable gloves) into the dish. Keep the yolk intact. When nearly all the white is separated, put the yolk sac over a small clean beaker and pierce it with a scalpel.

Allow the yolk to run out and discard the membrane. One egg can make quite a lot of paint.

Don't forget to clean up very thoroughly, finishing with disinfectant. Raw eggs can carry salmonella bacteria, which cause severe food poisoning.

Grinding the solid pigment. Do the initial grinding in a pestle and mortar. For some pigments, grinding to a very small particle size can affect the colour. To illustrate this, put a few crystals of hydrated copper(II) sulfate into a mortar and check the colour as grinding goes on.
NB Only use a small amount of pigment at a time.

Mixing pigment and medium to make paint. Experts say that it is much better to use a glass muller and slab for this job. However, a good pestle and mortar coupled with persistence and patience gives an acceptably homogeneous and workable paint.

After the preliminary grinding, put some powdered pigment in the mortar and add water a few drops at a time, mixing with a palette knife or spatula until a good slurry is obtained. Then add the egg yolk a drop at a time while mixing. The volume of egg used roughly equals the volume of water to get a useable paint; but this varies with different pigments. When the consistency seems right, grind the mixture thoroughly with the pestle in the mortar. Use good downward pressure, pausing often to scrape the paint-mixture off the pestle with the palette knife and putting it back in the mortar. Continue until the paint has a completely even consistency. Add a drop more egg yolk or a little more of the ground pigment if necessary. The paint should come freely off the brush but not be 'watery'.

Scrape the paint out of the mortar with the palette knife or a suitable spoon, and place it in a suitable small container ready for painting. It should be used very soon after being made. Do not make too much paint at a time. If a muller and slab are used, the instructions are basically the same. The mixture frequently needs to be scraped off the muller back on to the slab.

Mixing paints to make different colours. Some of the pigments are very intense and need to be mixed with white to obtain the desired effect. Or they can be mixed to obtain different colours – *eg* blue and yellow to make green. The proportions used are found by trial and error. Mixing can be done on any suitably smooth non-porous surface, using – *eg* a palette knife. Mixing must be done thoroughly.

Using your paint to make a picture

Teacher's notes   This can be the most satisfying part of the work!

Apparatus and equipment

- Beakers *etc*
- Hot water bath. (If an electric one is not available, a suitably sized beaker containing hot water will do.)
- Scissors
- Glass rods
- Spatulas
- Brushes of various sizes, including very fine
- Palette knife

Materials

- Wooden panels, square or rectangular, 30 cm edge or larger (but don't be too ambitious to start with!). Any wood, or wood composite, will do. One side should be reasonably smooth and flat. Use a straight edge to check, both ways
- Sandpaper, medium and fine; garnet paper (very fine – *eg* EAC 129) for finishing
- Charcoal sticks and/or a 4B pencil
- Gypsum (hydrated calcium sulfate) or fine precipitated calcium carbonate. Ordinary 'whiting' will do, but may need grinding
- Kitchen gelatine
- Paper towels
- Household aluminium foil

Cleaning   With both egg tempera and gesso, spills and waste can be wiped up with paper towels or tissues and the final traces washed away with hot water. Brushes used for egg tempera, glue or gesso should be rinsed **immediately** with plenty of water. Try not to let the materials harden before cleaning them up. (The mortar used to build the walls of the Kremlin in Moscow contains lots of egg. These walls have stood for many hundreds of years and should stand for many more.)

Student's instructions   *Preparing the panel.* The older paintings in this pack used wooden panels as a support. To start with, softwood or composite wood pieces about one foot square can be used. Use a straight edge to check for flatness, across as well as up-and-down! Rub with medium sand paper and wipe or brush clean; then 'size' the panel by brushing it with glue (see next page). Brush out any air bubbles. Allow to dry. Be careful not to inhale dust while sanding.

Making the glue. The proportion of gelatine to water varies depending on the source of gelatine. Sprinkle about 10 g of gelatine granules slowly on 100 cm^3 of cold deionised water. Heat gently, but do not boil. Stir gently. The liquid should be a uniform light yellow.

On cooling the mixture should set like jelly. If it does not, warm it again in a hot water bath to liquefy it and add a little more gelatine. If it sets too ' rubbery', warm to liquefy and add a little more water. If the mixture is not used all at once, a little more water may need to be added to keep the mixture at the right consistency. Use while at a comfortable hand heat.

Making the gesso. The older paintings in this pack used a gesso ground to make a smooth, hard surface for painting. Sprinkle the solid gypsum, calcium carbonate or whiting*, a little at a time on to the warm, liquid glue while it is in the beaker. Allow the solid to be absorbed and sink. Continue adding solid until no more sinks. While the mixture is still warm, stir gently, avoiding air bubbles in the mixture. The consistency should be similar to a creamy batter.

When cold, the gesso sets hard. The gesso can be liquefied by warming it in a hot water bath. It tends to settle and needs gentle stirring. If the liquefied mixture is too thick, add a little more water. As time goes by, more water may be needed anyway. Use while warm at a comfortable hand heat. The mixture should brush easily on to the panel surface.

Laying and finishing the gesso 'ground'. It will probably take four or even six coats of gesso to lay a good ground on your panel. The first coat should be brushed on, say, up-and-down, the next coat crossways, the next up-and-down, and so on. Each coat will take about 20–30 minutes to dry.

It is important that there are no air bubbles in the gesso. When each coat has been applied, the brush should be lightly drawn over the wet gesso the whole length or width of the panel to eliminate any air bubbles, however small. A little practice may be needed.

When enough coats of gesso have been applied, allow it to dry thoroughly. However carefully you have brushed, fine lines will be visible.

Sand carefully (beware dust!) with fine sandpaper, and finish with very fine garnet paper. Rub hard with tissue paper. The surface should now

---

* Whiting can be bought from ironmongers and is used for marking out sports pitches.

feel very smooth and hard. Finally, you may want to seal the gesso ground with a thin, uniform coat of glue (imprimatura) applied with a soft brush.

Preliminary drawing. It is better not to be too ambitious to start with! Decide on your design. Do you want any part to look metallic, or metallic showing through paint (**sgraffito**, see below)? The borders of these areas can be marked out by lightly incising the gesso with a stylus or scalpel. The remainder of the design can be marked out with charcoal or by drawing lightly with a soft (4B) graphite pencil. Show which colours are to go in which areas.

Applying the paint. The areas to be painted with different colours should be made clear on the design. The paint should be used, with suitable-sized brushes, very soon after it is mixed. Unlike oil paint, egg tempera dries quickly and cannot easily be worked and modelled. Effects of light and shadow, folds in clothing *etc* are usually achieved by painting on top of an already-dry layer.

You will find that the paint has to be applied with a short, light stroke, unlike the long strokes which can be used with oil.

NB Never lick a brush to get a good point.

Gilding, silvering and sgraffito

Ideally, gold leaf would be used to make marvellous gilding, such as is seen on old altarpieces including the one by Niardo di Cione which is the first painting in this pack. But gold leaf is expensive and difficult to work with. Imitation gold leaf ('Dutch metal') can be used, but it is very flimsy and hard to handle and discolours in a few weeks in the atmosphere of a school laboratory.

In genuine gilding, a layer of red-brown **bole** is often used to coat the gesso before the gold leaf is applied. Here, the metal is put directly on the sealed gesso ground.

If you cannot find gold coloured foil, household aluminium foil can be used to give a good silver effect, either by itself or through paint (**sgraffito**). It has a dull side and a shiny side. If you are trying to make your painting look old you should have the shiny side facing upwards.

Cut a suitable piece of foil (preferably a fairly small strip) and place it ready on a piece of firm white card with the end of the foil extending just beyond the edge of the card.

Wipe the area of sealed gesso to be silvered to make sure it is free of grit, then wet the area using a soft brush. Bring the 'loose' end of the aluminium to the wet surface, and hold it down lightly with a soft brush. Withdraw the card steadily and follow it with the brush until all the foil is on the surface. The foil can be gently manipulated on the wet surface. Brush it down. Dab with tissue if necessary. Trim with scissors or a scalpel. If a corner fails to stick, lift it, put a drop of water underneath, and brush it or pat it down gently. Other pieces of foil can be added and butted up to what is already there. Any gaps can be filled with small pieces of foil on the damp gesso.

When all the required areas are covered with foil, wipe it carefully with tissue or cloth to get rid of any particles of gesso. The foil surface probably looks quite rumpled. Using the rounded bottom of a test-tube (check that it is perfectly smooth) gently rub down any creases and ruffles while the underlying gesso is still damp. When it is flattened, and the gesso underneath is dry, the foil surface can be rubbed all over with the tube, gently at first and then a little harder, to polish it.

The technique of **sgraffito** was widely used, particularly in making altarpieces.

You can imitate it in this way. Decide on which areas of aluminium foil you wish to use, and choose a colour or colours. Paint these areas, using two or more coats if necessary until no trace of the underlying metal can be seen.

When the paint is 'dry' but not fully hardened, use a scalpel (CARE!) to gently scrape away the paint in a pattern to reveal the shiny metal underneath. Great care is needed – but the result can be very satisfying.

The old artists would then very carefully use a small hammer to tap a punch on the exposed metal to make a pattern of indentations – especially in haloes, which both looked good and gave extraordinary optical effects in candle light or lamplight. You may wish to try to do the same with aluminium foil.

Varnishing

Teacher's notes

Apparatus and equipment
- Small container
- Suitable brush

Materials
- Suitable varnish. (If none is available at school, it can easily be obtained from any shop which sells artists' materials.)

Student's instructions

When your panel is fully dry (which may take some weeks), you may wish to varnish it to coat and preserve its surface.

Carefully brush a thin layer of the varnish over all the surface of the picture. You can check that you have covered it all by holding the panel (without fingering the painted side!) so that light reflects off it at a small angle into your eye. Any 'unwetted' areas will clearly be visible.

Allow the varnish to dry.

Making and using oil paints

Teacher's notes

Apparatus and equipment

- Fume cupboard
- Pestle and mortar (or muller and slab)
- Selection of brushes, including fine and very fine
- Small containers for paints
- Palette knife
- Small plastic or wooden spoons
- Palette, or other surface on which paints can be mixed
- Small sand bath
- Bunsen burner, tripod, tongs, retort stand and clamp
- Thermometer (0–360°C)
- Paper towels

Materials

- Pigments, panel, gesso
- Linseed oil – raw, rather than a 'stand' oil
- Turpentine (not turpentine substitute). (**Hazard**: harmful, flammable)
- White spirit. (**Hazard**: flammable, toxic to aquatic organisms, harmful)
- Lead(II) oxide. (**Hazard**: toxic, may harm the unborn child)
- Small piece of muslin; length of thread

Cleaning

Do not mix more paint than you need. Most of the pigments here are intense in colour and a little can cover quite a large area.

Cleaning pestles, mortars and glassware – and spillages – should be done immediately, before the paint has begun to dry. Nearly all the paint can be removed using paper towels (wear disposable gloves) and disposed of normally or put in black plastic bags. The last traces of paint can be removed using a paper towel with a little white spirit.

Brushes should be thoroughly wiped with a paper towel, and then cleaned in the minimum of white spirit. (Your science department should have a standard method for disposing of the small amounts of contaminated organic solvents which result from normal teaching).

An explanation of the drying process for post–16 students.

> The 'drying' of these oils is a complicated process which is not fully understood. The oils themselves are triglycerides – *ie* three carboxylic acid chains are joined to a glycerol (propan-1,2,3-triol) molecule by ester linkages. The 'drying oils' have a characteristic carboxylic acid chain length of 18 carbon atoms, and the chains contain two or more carbon-carbon double bonds. The carboxylic acids in walnut and poppy oils mostly have two double bonds, and those in linseed oil mostly have three. The oils absorb oxygen, leading to a range of free radical chain reactions which eventually result in a crosslinked polymer. The liquid oil becomes a rubbery solid. Because of the wide variety of structures of individual oil molecules, the nature of the crosslinking reaction and the resulting three-dimensional structure are very complex.

Student's instructions

The use of oil paints gradually displaced the use of egg tempera because artists found they could manipulate and model them more easily.

Support and ground. You may wish to use canvas. However, it is probably cheaper and more convenient to use wood panels with a gesso ground as before.

Painting may be done direct on to the (ivory smooth!) ground, or the ground may be sealed with a thin coat of glue as for an egg tempera panel.

Preliminary drawing. This may be done with charcoal or a 4B pencil as before.

Pre-polymerisation ('heat-bodying') of linseed oil. The so-called 'drying oils', which were used most as a medium for paint, were poppy oil, walnut oil, and linseed oil. Of these, linseed is the cheapest and 'dries' the fastest, and we shall be using it here.

Cobalt, manganese, lead and some other metal compounds accelerate the 'drying' of the oils. Such compounds are called 'siccatives' – from the Latin siccus, dry. Small amounts of these compounds appear to react with and 'dissolve' in the oil.

To 'heat-body' linseed oil. (**Hazard**: linseed oil is flammable.) Put about 40 cm^3 of raw linseed oil into a 100 cm^3 conical flask. Suspend in the oil, using a thread, a small muslin bag containing about 0.2 g of lead(II) oxide

(PbO, **Hazard**: toxic). Place a thermometer in the oil, then put the flask and its contents in a sand bath over a Bunsen burner. Support the flask using a retort stand, with a clamp around the neck of the flask. Heat with a medium flame in an efficient fume cupboard. It is possible that strong overheating may ignite the oil vapour; using a conical flask minimises this risk. Do not let the temperature of the oil exceed 230°C. Try to keep the temperature between 200°C and 230°C. The oil starts to darken at about 200°C.

Heat for about 30 minutes at 200–230°C, but do not allow the oil to become very dark. Allow the flask to cool to room temperature, remove the muslin bag, and decant – *ie* carefully pour off the linseed oil (leaving any solid behind) into a stoppered and labelled small flask. You may wish to repeat this with one or more other samples, using a longer or shorter heating time.

The 'drying time' of different oil paint samples. For this, you need some lead white pigment (**Hazard**: toxic) (see p 6), raw linseed oil, your sample(s) of heat-bodied oil, turpentine (**Hazard**: harmful, flammable), and some space on a prepared panel with gesso ground, which has already been painted (using egg tempera or oil) in a colour other than white. The paint on the panel must be dry.

(a) To a little solid lead white add raw linseed oil a drop or two at a time in a pestle and mortar while grinding. Continue until a smooth paint is produced. Transfer the paint to a suitable labelled container, and clean the pestle and mortar.

(b) Repeat *(a)* using your sample(s) of heat-bodied oil.

(c) Mix a little raw linseed oil with an equal volume of turpentine. Use this mixture with lead white pigment as in *(a)*.

(d) Repeat *(c)* using your sample(s) of heat-bodied oil.

On your prepared panel, use the paints from *(a)*, *(b)*, *(c)* and *(d)* to paint dense white patches about 2 cm square. Make sure the patches are labelled!

Check the patches two or three times a day over the next few days, and note when each patch becomes touch-dry.

Questions

> (i)   Which dries quicker – the paint using the raw oil or the paint using the heat-bodied oil? If you used more than one sample of heat-bodied oil, does the length of heating affect the drying time?
>
> (ii)   Does the oil-turpentine paint mixture dry quicker than the mixture using only oil?
>
> (iii)   Does using a darker oil alter the colour of the patch?

Making the oil paints. Do not make too much paint at a time.

Do not make any paint unless you are going to use it within the next hour or two.

For each paint, put a suitable amount of solid pigment (note the hazards) in the mortar, and add a 1:1 mixture of heat-bodied linseed oil and turpentine (**Hazard**: harmful, flammable) one or two drops at a time while grinding with a pestle. Frequently scrape the pestle with a palette knife and put the material back in the mortar. Continue until a good, smooth, workable paint is formed. (If necessary, test it by using a brush to paint it on a surface). Transfer the paint into a suitable container using a plastic spoon or similar, and clean the pestle and mortar before mixing the next paint.

Underpainting. Many artists have used an underpaint – *ie* putting an undercoating on all or part of the area of the picture. For example, in his early townscape painting *'The Stonemason's Yard'* included in this pack, Canaletto used a grey underpaint below the sky and a yellow-brown one under the buildings. Underpaint can affect the overall appearance of the picture. If you decide to use an underpaint, let it dry before painting over it.

Making a picture. The panel and its ground are now ready; any preliminary drawing and/or underpainting has been done; the paints are mixed and waiting.

What happens next depends mostly on your creativity and artistic ability! A couple of points may help.

- For some parts of the painting, you may be able to use the paints just as you have made them. However, to get just the colour you want it is often necessary to mix two or more paints – *eg* shades of green, or to contrast highlights and shadows. Almost certainly you will need more lead white than any other of the pigments you have made. You will need a palette or a suitable surface on which to mix your colours.

- Because, unlike egg tempera, oil paints dry relatively slowly, it is possible to paint 'wet on wet', and work and manipulate the paint to get the effects you want. You may decide to build up the paint layer to develop an **impasto** effect.

You may choose to finish your picture in one go, or leave it for a while and come back to it. When the paint is fully dry, you may want to varnish it to protect its surface.

Demonstrating the principle of X-ray photography

Teacher's notes

Apparatus and equipment
- Overhead projector
- Pens.

Materials
- Greaseproof paper

Student's instructions

This demonstration can be done with pieces of greaseproof paper or with good tissue paper.

Draw or paint a simple design on a piece of the paper; cover it with blank pieces until the design is just no longer visible, then on top of it all place a final piece with another design on it. Now put the pile of sheets on an overhead projector lens and switch on.

NB You can't project the image; you have to look down on it.

Frequently asked questions

Why use the yolk rather than the white to make egg tempera paint?

When people realise that tempera paint is made with egg it often comes as a surprise to discover that it is the yolk (rather than the white) which is used. After all, yolk is often a strong yellow, while the white is virtually colourless. In fact the yellow colour of the yolk does not affect the colour of the paint at all. The reason for using the egg yolk is that the material in the yolk is in the form of an emulsion. An emulsion consists of two immiscible liquids apparently 'mixed' together; where droplets of one are 'suspended' in the other, they don't settle out into two layers as expected. An emulsifier is often used to enable an emulsion to form and stabilise. For example, mayonnaise is an emulsion made from vinegar and olive oil – with, oddly enough, egg yolk as the emulsifier.

Egg yolk consists of fatty compounds suspended in a mixture of proteins and water. (Egg white contains no fat, only the proteins – mostly albumen – mixed with water.) The egg fat plasticises the paint film, enabling it to dry to a semi-matt surface sheen without becoming brittle.

When egg tempera paint dries, water evaporates first. The egg proteins then 'set' to form a hard, waterproof film incorporating the pigment. In contrast to oil paint, which sets with only a small decrease in volume, evaporation of the water from egg tempera paint causes a considerable loss of volume. This is why egg tempera paint must be applied in successive thin layers, rather than thickly, to avoid cracking.

Because of its composition, egg yolk paint adheres to gold, enabling the technique of sgraffito to be used, as in the Nardo di Cione altarpiece in this pack. Artists do sometimes use egg white – *eg* in Elizabethan miniature paintings.

Why use calcium sulfate (gesso) or chalk to lay a ground?

The gesso (used in Italy) or chalk (used in northern Europe) was mixed with glue-size – *ie* glue diluted with water, to make a plaster very similar to what is used on walls and for making coving. The glue slows down the rate at which the plaster sets, and ensures that it is much harder once it is fully dry. The plaster ground is inflexible, which is why it has never been fully compatible with a canvas support, and can be scraped or sanded and finally buffed to produce an exceptionally smooth surface which is not too absorbent. (The ground can be coated with size, if necessary, before painting). Also, the whiteness of the gesso or chalk contributes to the brightness of egg tempera paintings, because the paint layers are rarely thick or opaque enough to hide the ground completely.

Why use cow hide, rabbit skins, bones etc to make glue?

'Glue' is the word used for adhesives which contain mostly gelatin, the important ingredient in jellies. Degrading collagen, the connective protein in animals, makes gelatin. Collagen is found in skin, cartilage, bone, tendons, ligaments and whatever else holds bodies together. In the disease called scurvy, which used to kill sailors by the shipload, connective tissue could not be regenerated and the body basically fell apart. Captain Cook found that lime-juice prevented scurvy, and eventually sailors were issued with the juice of citrus fruits along with their usual rum. We now know that it is the Vitamin C in fruits which prevents scurvy.

To make glue, the collagen-containing material is boiled slowly in water for hours. The resulting solution is then clarified and concentrated. Glues can be made from mammals or fish; those from fish are soluble at room temperature, but those from mammals need hot water. 'Skin glues' are usually stronger than the others.

As an alternative source, the protein casein can be obtained from milk. Casein can be made into an exceptionally strong adhesive by mixing it with a suspension of lime (calcium hydroxide).

As the gelatin in a glue 'sets', proteins in it form a strong three-dimensional crosslinked structure. (A jelly consists of water and flavourings and colourants contained in such a 'gel' structure).

All protein-based glues and adhesives are liable to microbial attack in damp conditions.

What are the advantages of gold leaf over silver leaf?

Silver leaf slowly tarnishes in air or water because it is more chemically reactive than gold. Traces of hydrogen sulfide or other sulfur-containing compounds in the atmosphere cause the formation of black silver sulfide. Where silver was used in paintings – *eg* for armour – it was often coated with lacquer or glaze; however, this isolation rarely prevented the blackening effetcs of air for long. Chlorides cause corrosion; and because silver chloride is light sensitive – remember silver halides are used for photography – this can also cause black stains. Gold on the other hand does not tarnish, and although more expensive was preferred. It can be beaten into uniquely thin but coherent sheets and burnished to give a brilliant shine.

The treatment of silver objects in museums and galleries seems to differ on a geographical basis. In western Europe, silver is lacquered or cleaned regularly to stop it forming black tarnish. In eastern Europe it is often deliberately left uncleaned and open to the air. Can you find arguments to support each approach to the problem?

The Chemistry of Art

The colour supplement

Introduction

This section is intended primarily for post-16 students. The material consists not only of chemistry and art but also physics and biology. However, much of the material is accessible to pre-16 students. Teachers can use this section as a source of facts and ideas.

How do we see colour?

The retina of the eye

The visual arts, such as painting, depend on vision – the ability to see.

A painter has to see while work is in progress, and the viewer has to see and respond to the painting. Because no two humans are the same, they probably do not see the same object in exactly the same way. For example, it is possible that the painter El Greco had a defect in his vision, which led to his unique style and characteristically elongated figures.

Colour is very important in vision. Colours are not visible in very dim light; there is a threshold of light intensity, varying from person to person, above which colour can be seen.

The process starts at the retina of the eye and results in nerve impulses to the visual cortex of the brain. The retina contains two kinds of receptor cells: about 100 million rods and 5 million cones. The rod-shaped cells are mostly found away from the centre of the retina. They can detect different levels of brightness, including extremely low levels of light intensity. Just five photons can be detected if they interact with five different rods. However, rods cannot detect colours. The cone-shaped cells mostly occur towards the centre of the retina, the *fovea*, and contain the pigments which make colour vision possible.

Except for primates and humans, mammals rarely have any colour vision. Pigeons possess only cones; they therefore have colour vision, but can only see in bright light. Owls have rods but no cones; so they are colour blind, but able to see in very dim light.

The chemistry of vision

When light strikes a receptor cell (a rod or a cone) chemical changes occur. These eventually generate an electrical signal in the optic nerve. The chemistry in rods is better understood than that in cones.

The pigment rhodopsin, 'visual purple', was isolated in 1878 from freshly dissected retinas. The red–purple colour in such retinas turns yellow and then colourless when exposed to light. In 1952 the structure which absorbs the light in rhodopsin was identified as the polyunsaturated aldehyde called 11-*cis*-retinal (*1*).

This aldehyde joins with a protein, opsin, to form rhodopsin; the aldehyde fits exactly into sites on the protein surface. The alternate double and single bonds in the retinal molecule enable rhodopsin to absorb light over a wide range of wavelengths.

(*1*)

When rhodopsin absorbs a photon, the 11-*cis*-retinal part is changed (isomerised) to the all-*trans* form of retinal (*2*).

This all-*trans* isomer is now a different shape and can no longer fit on to the opsin, and so detaches from it. It is this isomerisation and detachment that triggers the events which lead to a nerve impulse.

(*2*)

If light is shone at constant intensity into a live eye, rhodopsin is formed at the same rate as it is broken up. An equilibrium is established. If you are in darkness for 20–30 minutes, 'dark adaptation' occurs. The rhodopsin concentration in the retina reaches a maximum, and so therefore does the eye's sensitivity.

Rhodopsin can be re-formed in the retina both by strong light and by enzyme action.

3

The cycling of 11-cis-retinal in the body

11-*Cis*-retinal + Opsin $\xrightarrow{\text{light}}$ Rhodopsin

Visual purple (absorbs light, with maximum absorption at a wavelength of 498 nm)

Rhodopsin → All-*trans*-retinal + Opsin

Yellow (absorbs light, with maximum absorption at a wavelength of 387 nm)

All-*trans*-retinal + Retinal reductase → All-*trans*-vitamin A (3)

Colourless

All-*trans*-vitamin A (in liver) $\xrightarrow{\text{Enzymes}}$ 11-*Cis*-vitamin A

11-*Cis*-vitamin A (in eye) $\xrightarrow{\text{Oxidation}}$ 11-*Cis*-retinal

Vitamin A can also be formed in the liver from carotenes absorbed in the diet. Carotenes are found in nearly all green plants, and the orange colour of carrots is largely due to carotenes. The structure of β-carotene is shown in (4).

Oxidative breaking of the central double bond by liver enzymes gives two molecules of vitamin A.

Colour vision Thomas Young (a doctor at St George's Hospital, London) proposed the 'trichromatic' theory of colour vision in 1801. In this, the retina is considered to possess three kinds of colour receptor. The cells responsible for colour vision, called cones from their shape, were not directly observed until about 1960. The three sets of cones respond to different wavelengths. The 'green receptor' (G) is most affected by light of wavelengths near the middle of the visible spectrum; the 'red receptor' (R) by wavelengths near the red end; and the 'blue receptor' (B) by wavelengths near the (high energy, short wavelength) blue end. So visible light reaching the retina affects the R, G and B receptors. The sum of the different effects on each of the receptors, R + G + B, gives the brightness of the light; the ratio of R : G : B gives the colour.

4

Every colour, including white, excites the R, G, B receptors in its own unique ratio of R : G : B. So mixing red, green and blue lights in this same ratio should produce the original colour. Any colour can be matched by mixing the so-called 'primary colours' red, green and blue in the correct proportions. The great physicist James Clerk Maxwell demonstrated this in 1854. Since the rod cells can detect only variations in brightness but not colour, the pigments responsible for colour vision must be in the cones. However, the way in which the information from complex mixtures of wavelengths is sorted out and sent to the brain is complicated. There seem to be other cells in the retina which can analyse and compare the signals from the cones.

Colour blindness   Very few people are actually blind to colour, but many, especially white males, suffer from defects in colour vision. Colour blindness should be called 'colour defectiveness'. John Dalton, whose atomic theory of 1808 revolutionised chemistry, had colour defective vision and was among the first to study the subject. For this reason colour defectiveness was for many years known as 'daltonism'.

The usual defect is failure to distinguish between red and green. This is hereditary. About 8 per cent of white males of European origin have noticeably defective colour vision. The incidence of colour deficiency is low among Asians and Afro-Caribbeans.

The colour genes are found in the female sex chromosome (X). The defect in the gene is recessive – *ie* in the presence of a normal gene the person is not colour defective. This can be shown using a genetic diagram.

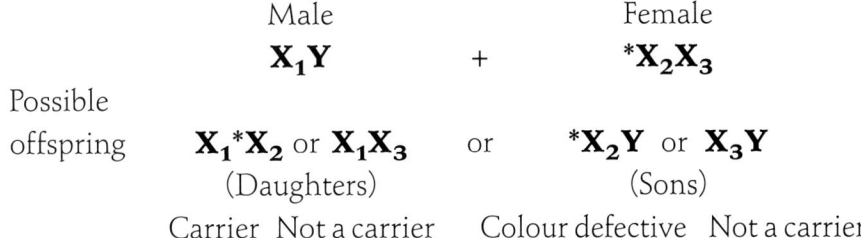

Male   Female

$\mathbf{X_1Y}$ + $\mathbf{{}^*X_2X_3}$

Possible offspring $\mathbf{X_1{}^*X_2}$ or $\mathbf{X_1X_3}$ or $\mathbf{{}^*X_2Y}$ or $\mathbf{X_3Y}$

(Daughters)   (Sons)

Carrier   Not a carrier   Colour defective   Not a carrier

However, if both parents possess the defective gene their daughter can be colour defective. The incidence of colour defective vision in females is less than one per cent.

People who have no cone pigments at all can see only black and white and shades of grey. This very rare condition is called 'monochromatism'. More commonly, one of the three sets of cones does not function. This condition is 'dichromatism', and leads usually to problems in distinguishing between red and green. Reflection techniques have shown that instead of having the cone pigments sensitive to both red *and* green, dichromats have only red *or* green. A lack of the blue-sensitive pigment is very rare. However, most people with colour defective vision are only partly deficient in one of the pigments. This could be because the absorption peak of the pigment is displaced from its usual wavelength, perhaps because the molecular structure is subtly different from that of the normal pigment. Mild colour defectiveness need not diminish the enjoyment of paintings, or affect the quality of life in other ways. However, if the condition is severe the person concerned will not be able to do certain jobs involving colour perception – *eg* train driving.

Colour defectiveness is easily detected and analysed using a set of charts designed by the Japanese ophthalmologist Ishihara. Each chart consists of dots of various colours and sizes. Those with 'normal' vision see a certain number or pattern; those with colour defective vision do not see some of these patterns, or see something different which those with normal vision cannot see.

The visible spectrum

Only a tiny part of the electromagnetic spectrum is visible to humans. The human retina can respond to violet light at a wavelength of about 420 nm. Radiation of rather shorter wavelength, and therefore of higher frequency and energy, is called ultraviolet (UV) radiation. Radiation from the sun contains ultraviolet; the more harmful shorter wavelengths are filtered out by the atmosphere, and especially by the small amounts of ozone in the upper atmosphere. Even so, at higher altitudes the retina can rapidly become affected (hence the need for goggles when skiing or climbing high mountains). Any breakdown of the so-called 'ozone layer' – *eg* caused by diffusion of CFC aerosol propellants into the upper atmosphere – leads to an increase in skin cancer and other undesirable biological effects. These biological effects might include destruction of the phytoplankton in surface layers of the oceans, leading to a significant reduction in photosynthesis and to an acceleration in global warming.

The upper wavelength limit for the human retina is about 700 nm – corresponding to red. Radiation of longer wavelengths, and therefore of lower frequency and energy, is called infrared (IR) radiation. The greatest heating effect of radiation is found in the infrared region.

Both ultraviolet and infrared radiation are widely used in chemical analysis. Infrared is particularly useful for studying the underdrawing of paintings. Infrared radiation can penetrate varnish and surface layers of paint better than visible light can, and is strongly absorbed by graphite pencil marks or by black inks which use carbon. The visible spectrum is revealed when 'white' light is passed through a prism or undergoes total internal reflection in raindrops – *ie* in a rainbow.

Mixing colours

Three words that are widely used in discussing colour are brightness, hue and saturation.

Brightness (also known as *intensity* or *value*). This depends only on the intensity of the light. Light of any colour can vary in brightness.

Hue. This refers to what we usually call the 'colour' of the light – *ie* green, red *etc.*

Saturation (also known as *tone* or *chroma*). This refers to the relative purity of the hue. A 'pure' colour, such as the blue of ultramarine or the red of vermilion, mixed with white gives a paler, unsaturated colour.

The eye can distinguish about 10 million colours. These colours can be mixed in two ways, additive and subtractive. The **additive primary colours** are red, green and blue. Almost every colour can be matched by mixing red, green and blue light beams in different amounts. If they are mixed in equal amounts, white is the result.

Subtractive colour mixing can be demonstrated by putting filters in a beam of white light. A red filter absorbs wavelengths corresponding to all colours other than red; only red is transmitted through the filter. Similarly, a green filter transmits only green. No light can get through if a red and a green filter are used together.

Pigments and dyes are also coloured by a subtractive process, but this time the colours that are not absorbed by the material are reflected. A blue pigment absorbs mostly yellow, red and orange while reflecting blue and violet. A yellow pigment absorbs blue and violet and reflects yellow, green and red. So, if yellow and blue pigments are mixed, the mixture is green, because this is the only colour which is not absorbed by either pigment.

The **subtractive primary colours** are obtained when the additive primary colours are subtracted from white light.

> White minus red leaves blue and green – *ie* cyan.
> White minus green leaves blue and red – *ie* magenta.
> White minus blue leaves green and red – *ie* yellow.

So the subtractive primary colours are cyan, magenta and yellow. These are the three colours which are used in colour printing, and also in some processes for colour photography – *eg* in a process called dye diffusion thermal transfer, which is used to make colour prints from electronic cameras.

7

What makes coloured objects coloured?

An object is visible because some light is reflected from it into the eye. If all the wavelengths in sunlight are reflected, the object is white; if all are absorbed, it is black. If some of the visible wavelengths are absorbed and others reflected, the object is coloured.

Some solid materials, such as glass, are transparent. Ordinary glass, which contains mostly sodium, silicon and oxygen, is colourless. Adding certain materials – *eg* cobalt compounds or colloidal gold – to molten glass can result in beautiful 'stained glass', blue and red respectively. Here, it is the transmitted light that is coloured – *ie* the light that passes through the glass. Note that stained glass is transparent but not colourless. Using the word *clear* as a synonym for *colourless* is wrong – *eg* when used on glass recycling skips.

Pigments and dyes

The main difference between pigments and dyes is that pigments are insoluble in egg yolk, linseed oil, water or whatever medium is used, and dyes are soluble. Clothes and hair can be dyed, but paint requires pigments. Pigments can be mixed with a medium and spread on a surface, as in painting or printing, or incorporated into the material from which the object is made – *eg* a plastic washing-up bowl. Dyes attach themselves to the molecules of the material to be coloured, and are therefore always distributed throughout the bulk of the coloured object.

Most dyes consist of organic molecules which contain delocalised aromatic systems, whereas most pigments are inorganic. Traditional 'lake pigments', which were used in several of the paintings in this pack, consist of natural dyestuffs, such as bright red cochineal (5) obtained from scale insects. These are adsorbed strongly by hydrated alumina. By sticking so strongly to the insoluble alumina the dye is itself made insoluble, and so can be used as a pigment.

(5)

Complementary colours

When sunlight is shone on to an object and light of a particular wavelength is absorbed, the colour corresponding to that wavelength is subtracted from the white light. The remaining wavelengths are reflected, and the 'complementary colour' is seen. This effect can be summarised using a colour wheel (*Fig 1*).

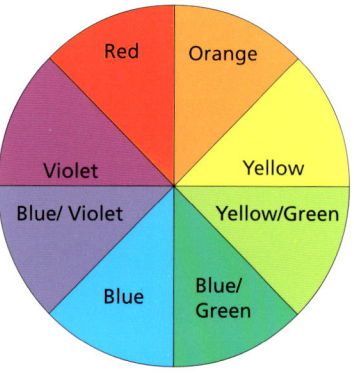

Fig 1. A colour wheel.

8

The colour wheel shows that an object that absorbs yellow light appears blue-violet and an object that absorbs orange light appears blue. So vermilion is red because it absorbs blue-green light, and ultramarine is blue because it absorbs orange light.

The colour wheel may give the impression that the different colours occupy equal portions of the visible spectrum. This is not the case. *Figure 2* shows that the red or blue regions of the spectrum occupy a far greater band of wavelengths than – *eg* the orange region.

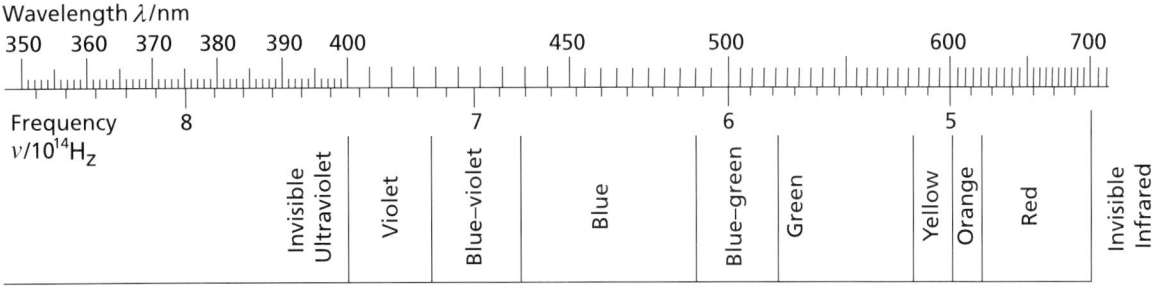

Fig 2. The electromagnetic specrtrum.

Reflectance spectra   The absorption spectrum of a material can be obtained after dissolving it in a suitable colourless solvent. But a pigment in paint is, by definition, insoluble. However, a pigment can be studied by shining white light on a part of the surface of the painting which contains the pigment and then examining the spectrum of the reflected light. This technique can be used in examining paintings.

For a solution of a blue material, red and yellow light is absorbed and blue light is **transmitted**. For a blue pigment in a painted surface, red and yellow are again absorbed and blue light is **reflected**. *Figure 3* shows the absorption spectrum and reflectance spectrum of the blue pigment Monastral Blue.

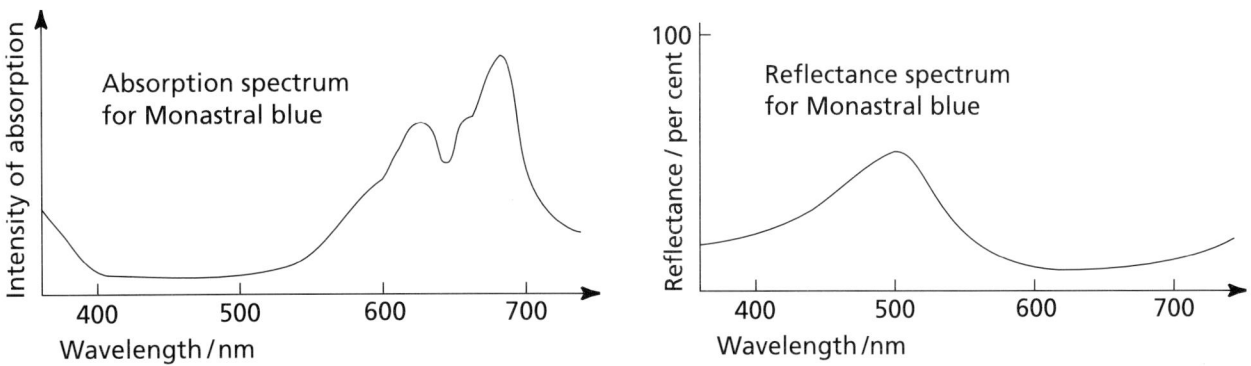

Fig 3. The absorption spectrum and reflectance spectrum of Monastral blue.

Realgar mineral

Cinnabar

Red ochre

Yellow ochre

Natural realgar from mineral

Synthetic vermilion

Burnt umber

Green earth

Orpiment mineral

Black

Iron oxide (red)

Natural orpiment from mineral

Pigments available up to 1600 AD

Lead white · Lapis lazuli mineral · Azurite mineral · Malachite mineral

Red lead · Natural ultramarine · Natural azurite (low grade) · Natural malachite from mineral

Lead-tin yellow (Type I) · Blue glass for smalt · Natural azurite (high grade) · Synthetic malachite

Lead-tin yellow (Type II) · Smalt · Synthetic azurite · Verdigris

Before chemists got to work in the 19th century the range of pigments available to artists was limited. Artists' colours were obtained from three main sources: minerals, plant/animal, and coloured earths. This chart shows the majority of the colours available and some of their sources.

17th and 18th century pigments

Kermes insects

Cochineal insects

Madder root

Kermes lake

Cochineal insects enlarged

Madder lake

Naples yellow (reddish)

Stick lac

Cochineal lake

Indigo

Prussian blue

Lac lake

Yellow lake

Scheele's green

13

Figure 4 shows the reflectance spectra of some other blue pigments. It is easy to distinguish many pigments using this non-destructive technique.

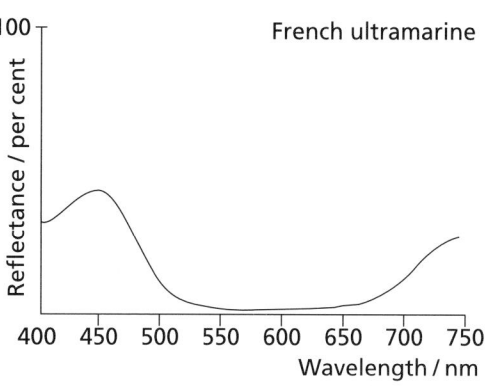

The causes of colour The majority of pigments and dyes commonly used consist either of compounds of metals from one part of the Periodic Table (the d-block), or of quite large organic molecules. Two famous and intense pigments, vermilion (red) and ultramarine (blue), fall into neither category; nor do the **earth colours** and red lead. The reasons for their colour are discussed later (p 21).

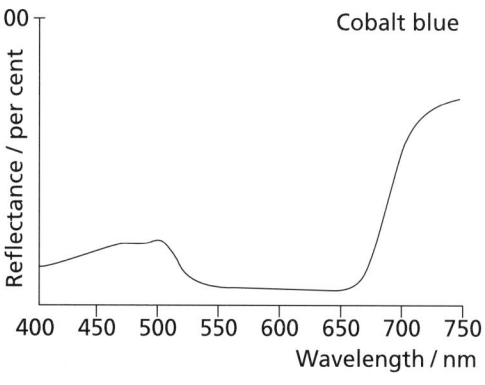

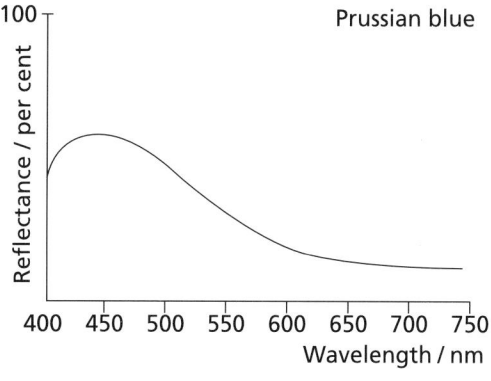

Fig 4. *The reflectance spectra of some blue pigments.*

When matter absorbs light energy as photons, the light is transformed into other forms of energy. Those photons that have energies corresponding to the energy differences between possible electron levels are absorbed. $\Delta E = h v$ gives the energy difference, where h is Planck's constant and v is the frequency of the radiation.

The excited electrons are the outermost ones in the atom, molecule or ion. Inner electrons are held more strongly. To excite them requires more energy than visible or most ultraviolet light can provide. (The production of X-rays involves electrons close to the nucleus of large atoms such as copper.)

The excited electrons can fall back to the lowest energy level – the ground state level – either directly or via intermediate energy levels.

As they do so photons of light energy are given out. The energy and the frequency of the emitted light are determined by the energy difference between two levels. So if the return to the ground state does occur via an intermediate energy level, the light emitted has a lower energy, and therefore lower frequency and longer wavelength, than the original light which caused the excitation. This effect is called **fluorescence**.
For example, if a ruby is placed in invisible ultraviolet light, it fluoresces bright red. The photons of the fluorescence are at the low energy – *ie* long wavelength – end of the spectrum.

In most coloured materials the colour is due to some wavelengths of the incident light being absorbed through exciting electrons into higher energy levels. However, when such electrons fall back to the ground state, the energy is released as heat rather than light.

Electronic transitions are always accompanied by smaller changes in vibrational and rotational energies. One effect of the absorption of light is that some of the light energy is transformed into vibrational and rotational energy and the object becomes slightly warmer.

Colour in organic compounds

Visible light is of quite low energy, so the gaps between the electron energy levels involved in the absorption and emission of visible light are quite small. In organic compounds these small energy gaps exist in structures containing either benzene rings, or several alternating carbon-carbon single and double bonds – this is called **conjugation**.

Cochineal (5) is an example of a dye containing benzene rings, β-carotene (4) is an example of a coloured natural material containing a conjugated system. Some of the electrons in a conjugated system are delocalised – *ie* they are not fixed in their position, as they would be in a normal covalent bond.

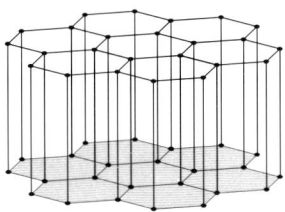

Fig 5. The structure of graphite.

The ultimate example of delocalisation in carbon-based structures is graphite. In this form of carbon, six-membered rings of carbon atoms are joined together to form large planes of atoms (*Fig* 5) in which one electron from each carbon atom is totally delocalised. This is why graphite is a good conductor of electricity. Because of this delocalisation, the energy levels in graphite are very close together. So graphite is a very good absorber of radiation, including all visible wavelengths, and is therefore black. It absorbs infrared radiation strongly, which is how infrared radiation is used to detect underdrawing in paintings.

All organic dyes or pigments contain delocalised systems. Examples of natural dyes, obtained from plants, which are used in making lake pigments are indigo (*6*, blue) and alizarin (*7*, red).

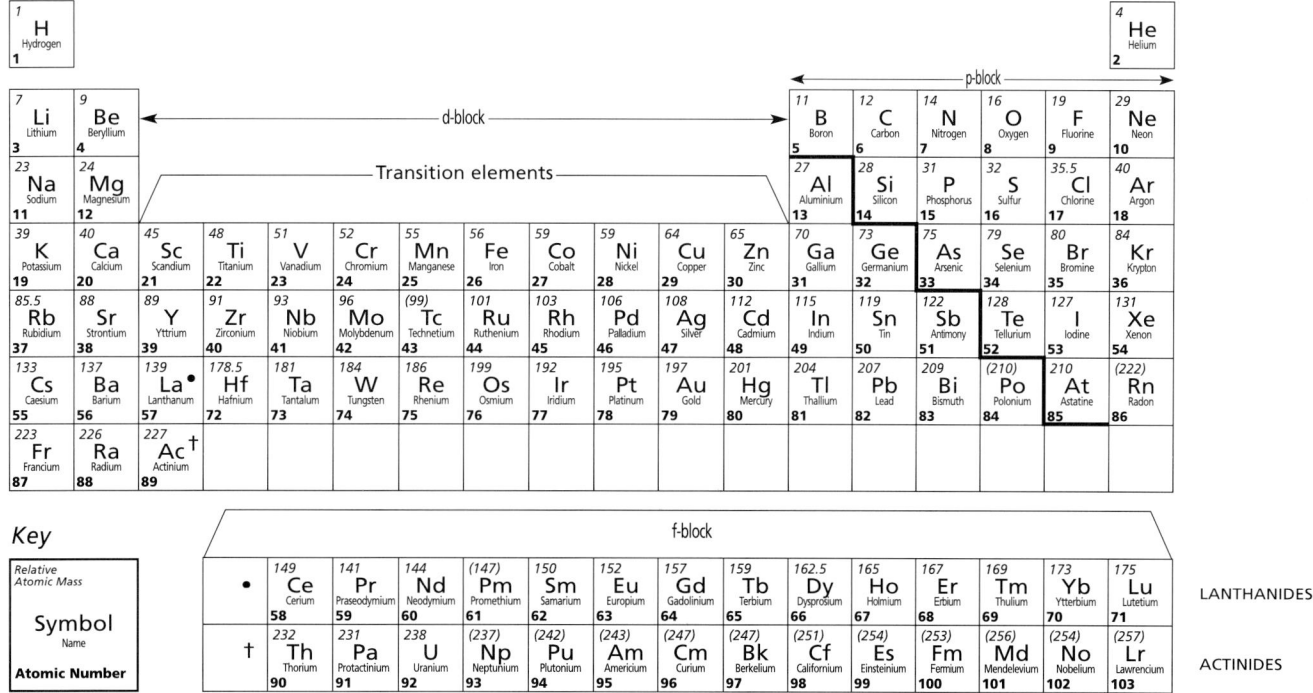

(*6*)   (*7*)

Delocalised systems are essential in the two most important pigments of all – chlorophyll, which traps sunlight in the first stage of providing energy for almost every living thing, and haemoglobin, which carries oxygen around our bodies.

Colour in inorganic compounds

The atomic number of an element is equal to the number of electrons in each atom of the element. These electrons are found in energy levels in atoms, and each main energy level has one or more sub-levels associated with it. These sub-levels are labelled s, p, d, and f. The s sub-level can hold only two electrons, the p sub-level six electrons, the d sub-level 10 electrons and the f sub-level 14 electrons. Each sub-level consists of electron orbitals, each of which can hold a maximum of two electrons. So the s sub-level has only one orbital, the p has three orbitals and the d has five orbitals. The Periodic Table (*Fig 6*) shows this 2, 6, 10, 14 pattern.

Fig 6. The Periodic Table of the elements.

The d-block of the Periodic Table (still sometimes called the 'transition elements') contains many of the metals which are found in pigments – *eg* chromium, manganese, iron, cobalt, copper, cadmium and mercury. These metals lack a full, stable, outer energy level of electrons.

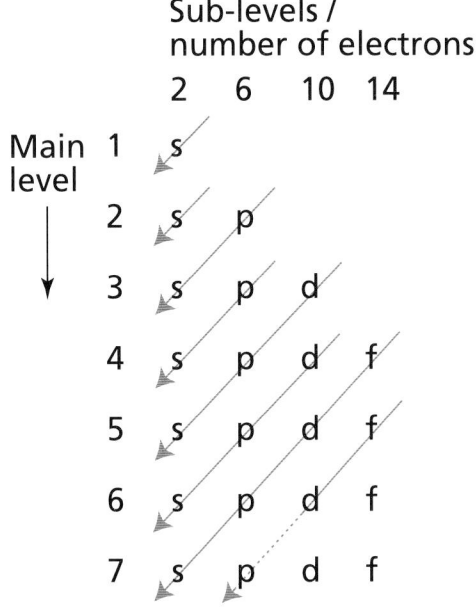

Fig 7. The order of filling of electron energy sub-levels.

The order of filling of the sub-levels is shown in *Fig 7*. Electrons fill the levels of lowest energy. This diagram gives the correct electron structure for all but a very few elements – *eg* chromium and copper. Even then, the difference only involves one electron switched into the next sub-level up or down.

In the first row of d-block elements the 4s, 3d, and 4p sub-levels are all close together in terms of energy. With organic dyes, when electrons can jump between energy levels which are close together, visible light is absorbed or emitted. The material is coloured. d-Block elements can also form complexes in which the energy levels are also close together.

In a complex, a central metal atom or ion is surrounded by negative ions or neutral molecules called ligands. Each ligand donates one or more pairs of electrons into vacant orbitals in the metal atom or ion.

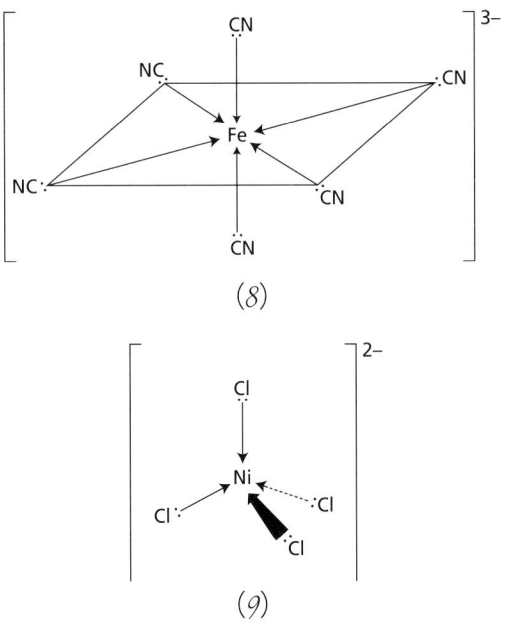

(8)

(9)

The total number of bonds (donated electron pairs) from the ligands to the central atom or ion is the coordination number in the complex. The most common coordination numbers are four and six. A coordination number of six almost always means an octahedral complex – *eg* (8). Two shapes are possible for a coordination number of four, tetrahedral (9) and square planar (10).

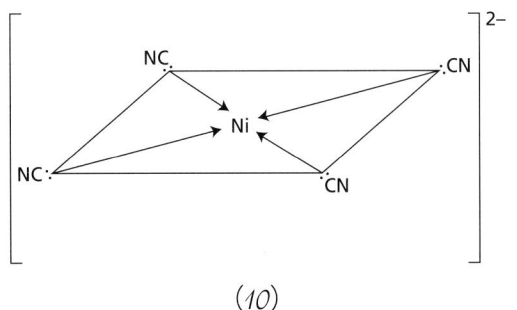

(10)

Where there are no ligands the five d-orbitals in an atom of a d-block element all have the same energy. However, ligands can have an effect on the d-orbitals. Those orbitals that are close to the ligands are raised in energy; the others become slightly lower. This d-orbital splitting is caused by the *ligand field effect*.

19

For many complexes this energy difference is in the visible region of the spectrum, so they are coloured. The colour of the complex depends on the number of d-electrons, the geometry of the ligand arrangement around the metal, and what the ligands are.

This d-orbital splitting by ligands is largely responsible for the colour of several well-known pigments. Examples include the green–blue copper pigments malachite, azurite, and the verditers; the green chromium pigment viridian; cobalt blue; and the pigment made from blue cobalt glass – smalt.

Most pure chemical compounds are colourless. In these compounds all the outer electrons of the atoms or ions are paired and need very high energies to excite them to higher energy levels and/or unpair them. So visible light, which has low energy, has no effect. This is also seen in the element carbon, in the form of diamond. All the outer electrons of the carbon atoms in diamond are paired and in strong covalent bonds. Diamond is therefore colourless, unlike the other form of carbon, graphite, which is black (see p 17).

Aluminium oxide is colourless when it is pure. Each aluminium atom is surrounded by six oxygen ligands in a roughly octahedral arrangement. But if the aluminium oxide contains a small amount of chromium impurity, so that perhaps one per cent of the aluminium atoms are replaced by chromium, a dramatic change occurs. The chromium has unpaired electrons; the ligands split its energy levels; light is absorbed at the blue end of the spectrum, and the result is a beautiful red transparent material which is known as ruby. Similarly, replacing some of the aluminium atoms of beryllium aluminium silicate with chromium results in emerald. The colours of many other gemstones are caused by ligand splitting of energy levels in small amounts of d-block elements which are present as impurities.

Colour due to charge transfer

Aluminium oxide which contains about 0.03 per cent titanium is colourless. If it contains that amount of iron, it may be pale yellow. But if both impurities are present, the result is sapphire, a lovely deep blue gemstone.

If sapphire absorbs a photon of the correct energy, an electron is made to shift from one metal ion to another. This results in a very temporary change in the charge on the ions.

$$Fe^{2+} + Ti^{4+} \rightarrow Fe^{3+} + Ti^{3+}$$

The photons absorbed correspond to the yellow–orange part of the spectrum, so blue, the complementary colour, is seen.

This charge transfer effect is the cause of the colour in Prussian blue, chrome yellow and chrome orange, all of which contain d-block elements; and the magnificent blue and very ancient pigment ultramarine, which does not. Ultramarine is a sodium aluminosilicate containing sulfur impurities. Charge-transfer processes involving sulfur are responsible for the colour.

Semiconductors and band theory

To understand why some common pigments are coloured and are also semiconductors, we need to introduce band theory.

In a metal, the outer electrons of the atoms are able to move throughout the metal; they are fully delocalised. Even so, the electrons occupy definite energy levels, but there are many of these, and in a metal they form a band of energies.

In many substances, which are not metals, a gap appears in the band, resulting in two bands, a lower valence band and an upper conduction band. The lowest-energy radiation which can be absorbed has to be able to excite an electron from the top of the valence band into the bottom of the conduction band. This energy is known as the *band-gap energy*. Any radiation which has energy higher than this can be absorbed and can excite electrons.

The band-gap energy in diamond is well into the ultraviolet part of the spectrum, so diamond is colourless. A material with such a large band-gap energy is a very good insulator because the transfer of electrons across the gap is more difficult.

Cadmium yellow (cadmium sulfide) has a band-gap energy of 2.6 eV, corresponding to light of wavelength about 480 nm; this means that it absorbs in the blue–violet part of the spectrum, so what is seen is the complementary colour yellow. The dramatic red of vermilion (mercury

sulfide) is caused by its small band-gap energy (2.0 eV), corresponding to about 620 nm. This means that it absorbs all light of wavelengths shorter than 620 nm, which leaves only red. If the band-gap energy is less than 1.77 eV, which corresponds to about 700 nm, all light is absorbed and the material is black – *eg* galena (lead (II) sulfide).

Most of the modern information and communications technology industry, including communications, timekeeping, control of machinery and the direct conversion of sunlight to electricity, depends on the manipulation of band-gap energies in slices and layers of slightly impure silicon.

Some other causes of colour

Scattering. This occurs when light shines on irregular surfaces or through fine suspended particles. The light is reflected in all directions. When the particle size is very small compared with the wavelength of light, scattering increases very rapidly with decreasing wavelength. Blue light is scattered much more than red light.

Sunlight is scattered by dust particles and random clusters of air molecules on its way through the atmosphere. The sky is blue because the intensely scattered blue radiation is seen against the black of Space. When the sun is very low in the sky at sunset or sunrise, sunlight has a much longer path through the atmosphere to the observer's eye. The blue radiation is scattered to such an extent that colours at the red end of the spectrum are seen.

Scattering produces the colour of most blue and green feathers, and is also responsible for the blue of blue eyes.

As the size of the particles responsible for scattering increases to, or becomes greater than, the wavelength of the incident light, so colours other than blue are seen. This is probably why colloidal gold was used in producing the superb red colour in medieval stained glass.

Interference. This occurs when light waves reinforce or cancel each other. Light waves are reinforced when peaks of waves coincide. Cancellation occurs when the peak of one wave coincides with the trough of another. If a beam of light falls on a thin transparent film, some of the beam is reflected from the top surface of the film and some from the back. The two reflected beams may reinforce or cancel each other. If white light is shone on the film, the overlapping interference bands from the different wavelengths give Newton's colours – *ie* the colours of the spectrum that are seen, for example, in a thin oil slick on a puddle. The colours can also be seen in soap bubbles and the coatings on camera lenses.

Biological structures which contain many thin layers also show this effect. These include peacock feathers, dragonfly wings, and the scales on beetles. If the layers are backed by the biological pigment melanin, unreflected light is absorbed. This causes iridescence – *ie* the colour looks metallic and changes as the angle of light changes. Some animals have such structures in their eyes. This is why their eyes appear coloured when caught in the beam of a torch or headlight.

Diffraction. This is caused by interference occurring at an edge. A diffraction grating is formed from a regular arrangement of edges or grooves in two or three dimensions, where the distance between repeating units is comparable to the wavelengths of light. Such a grating scatters light. As the scattered waves mix, they may reinforce or cancel, and in some directions they give the colours of the spectrum. The surface of a compact disc demonstrates this effect dramatically in ordinary daylight. The surface of a vinyl record, held so that light from a light bulb strikes it at a shallow angle and is reflected into the eye, gives two complete spectra and part of a third.

Diffraction effects are responsible for the colours on some snakes and the wings of some beetles. The precious stone opal contains a three-dimensional regular arrangement of tiny spheres of diameter about 250 nm. It is diffraction caused by this array which causes opal's characteristic appearance.

Practical activity

Demonstrating the ligand field effect

Add dilute ammonia solution dropwise to blue copper(II) sulfate solution, with stirring.

Note the various colour changes.

What do you think the pale blue precipitate is?

Explanation

The complex in the original blue solution is the octahedral $[Cu(H_2O)_6]^{2+}$ ion. When ammonia is added, the initial pale blue precipitate is copper(II) hydroxide. This is usually written as $Cu(OH)_2$, but is more likely to be $[Cu(H_2O)_4(OH)_2]$ in which the negative hydroxide ions have neutralised the positive copper ions so that the neutral particles can aggregate into a precipitate. The hydroxide ions are present in quite high concentration in the ammonia solution because of the equilibrium:

$$NH_3\,(aq) + H_2O\,(l) \rightleftharpoons NH_4^+\,(aq) + OH^-\,(aq)$$

The ammonia solution is therefore alkaline. Hydroxide ligands, like water ligands, can be pushed out when the concentration of ammonia molecules is high enough. The hydroxide ions are stronger ligands than water molecules, but the ammonia ligands are stronger still and the deep blue-violet $[Cu(NH_3)_4(H_2O)_2]^{2+}$ is formed. The colour change happens because ammonia ligands split the d-orbitals more than water ligands, so that ΔE is larger. The absorption spectra (*Fig 8*) show what happens. The ordinary hydrated ion absorbs red and appears blue. The ammonia complex absorbs mostly in the yellow region, and so it appears to have the complementary colour, blue-violet.

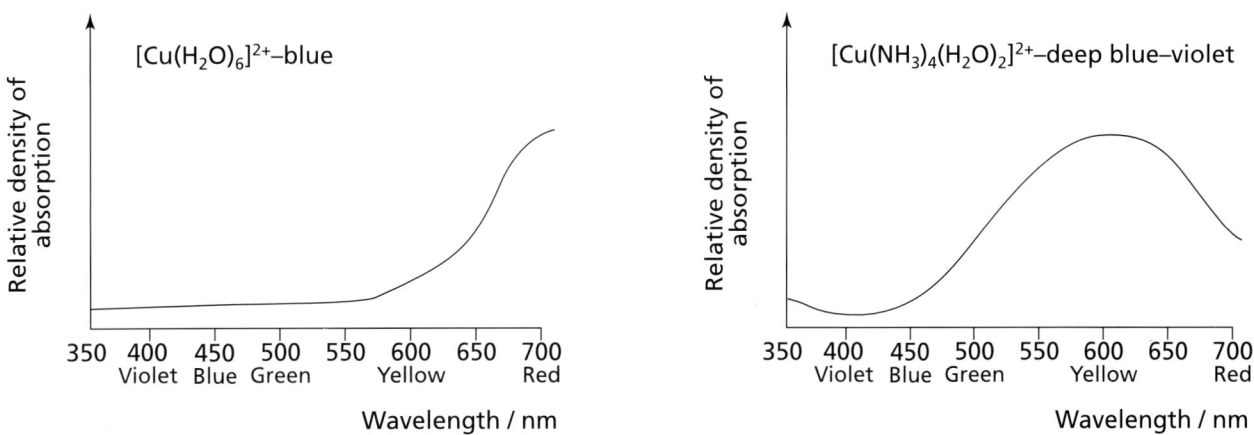

Fig 8. The absorption spectra of two copper complexes.